D1648853

HOW THE EXPERTS BUY AND SELL GOLD BULLION, GOLD STOCKS & GOLD COINS

HOW THE EXPERTS BUY AND SELL GOLD BULLION, GOLD STOCKS & GOLD COINS

JAMES E. SINCLAIR
AND HARRY D. SCHULTZ

ARLINGTON HOUSE · PUBLISHERS
NEW ROCHELLE, NEW YORK

Manufactured in the United States of America

Library of Congress Cataloging in Publication Data

Sinclair, James E
How the experts buy and sell gold bullion, stocks, and coins.

1. Gold. 2. Investments. 3. Hedging (Finance)
I. Schultz, Harry D., joint author. II. Title.
HG293.S54 332.6'3 75-19110
ISBN 0-87000-309-7

Dedicated to the hopeful expectation that
man will rediscover the principles of
fraternity in the days ahead.

Contents

Introduction

A Devil's Advocate on Gold

Amsterdam

People tend to be fanatic about gold. They either foam at the mouth in hatred of it or are so blindly progold that they begin mistaking the virtue of the gold mechanism for an all-wise, all-powerful Godlike force.

A sane middle ground would make for a more balanced stand that, in turn, can protect an investor better, make more profits, and give the objectivity that is necessary—since you should sell gold at times. In other words, gold should be treated like a stock or commodity instead of a religion.

This may appear to be a surprising position for me to take, for I am known as a "gold bug." But it is most important, in my opinion, to keep gold in its proper perspective.

I would be the first person to celebrate if some government came out with a plan that would completely abolish gold from the monetary system—that is a plan that would work—one that provides discipline over spending in an impartial way. But no such plan exists. It's unlikely one will ever be devised; therefore we are stuck with the metal. So let's examine what's wrong with gold.

One obvious flaw is that, as a medium of exchange, it is very heavy and cumbersome. You can't put a million dollars of gold in your pocket; it weighs a ton and takes up a lot of room. You can, however, put a million dollars worth of shares, or a letter of credit, or a check book in your pocket.

Pure gold is too soft to be used as coinage; it would bend and literally wear away. Perhaps that would be the ultimate in inflation—where we all had 100-percent pure gold coins and nothing else, and they just got smaller and smaller through wear till the gold supplies of the world disappeared. You HAVE to mix gold with other metals to be able to use it in any real sense. While gold bars are made of nearly pure gold, people don't and can't hand those around for their weekly groceries, or even for car purchases.

Gold thus has little practical use in bulk. Yes, it is used for some types of glass windows to preserve heat, it is used for certain areas of space travel and for electronics, but these are fringe uses. Its main use is to be looked at. We hang it on ourselves, stand it in our homes, or hide it in vaults and tell ourselves it represents wealth when we look at it.

But we don't and can't really USE it.

It is also very expensive to produce. One must process an enormous amount of rock, dug in the most horrendous underground conditions, to extract this yellow metal, ounce by ounce. And for what? For a metal whose biggest use is to be looked at?

People rob for it, people die for it. Governments overthrow whole institutions for it. Henry VIII dissolved the monasteries in England and claimed he had a new religion; partly because he wanted to take a new wife, but mainly because the country needed wealth. With the monasteries dissolved, their gold holdings and priceless altar vessels and ornaments could go into the government treasury.

Isn't it all rather silly that in 1975, with complex computers, with men rocketing to the moon, with instant global electronic communication, in such a seemingly advanced world a man's word cannot be sufficient backing for his international trading?

I wish it could be so. How much time and energy are spent, pain and misery are caused because man is still barbaric. He must use a silent, nearly useless, master, a metal, as a way of forcing himself to keep his word when trading with others.

But won't gold become outdated? Surely at this stage of civilization we can look into the very near future and see that something less barbaric than gold can govern our international trading. But what is there to choose? NATO? The UN? The IMF has proved it can't do the job. At this point in time it would seem that no man or body of men is capable of controlling money honestly and fairly to ensure its international stability.

Man just isn't honest. It is a terrible admission, but he isn't. Governments have brought the Western world to the brink of economic disaster where many of us live in fear of losing everything we have. Prices of goods and services have increased a hundredfold in a period when the amount a single man could produce has also increased a hundredfold. The equation here just doesn't come out right. When a man can produce ten times more per man-hour than he could say fifty years ago, prices should be much less than they were fifty years ago—this would be logical. But governments stepped in. They meddled with money. Yet we sit back and praise the politicians who say, "You have never had it so good, look what we have done for you."

What they should say is, "You have never had it so good. Look how the intelligent among you, the inventors and entrepreneurs, have raised your standards, and even we in Government, with all the grandiose plans that we made you pay for, couldn't take away all the effects of honest-to-goodness inventive progress."

And so we come back to gold. Not because we believe in gold. Gold is more a devil than a God. Gold is a monument to man's barbarity, not to his highest self. It is a reminder that without some sort of mutual economic weapon that everyone can use, man's inhumanity to man, man's greed, man's lust for power and fame will cause him to lie, cheat, steal, distort, and, if he is a politician, to do so with a smiling face, saying it is for your own good and in the public interest.

Gold is a weapon as primitive as the bow and arrow. To date nobody has invented economic gunpowder, although perhaps one day somebody will. Like atomic energy, suddenly a new concept of energy, *economic* energy, will break forth and start a new era. The splitting of the atom launched us technologically into the space age, but the rest of the departments of our life haven't caught up with that. Emotionally, economically, and in our relations to one another we are still in the stone age, and until we can break into the space age in these areas, we are stuck with a stone-age medium of exchange.

Look at the munitions business. Society has always looked down on these "merchants of death" as they have so often been labeled. People get emotional about it from time to time. Remember the antagonism toward the Krupp works? How students have demonstrated outside the factories of DuPont and similar manufacturers of war material? But the munitions business is booming, if you'll pardon such a low pun.

Munitions sales have increased 100 percent in the last ten years. The world has doubled its total armament in the past decade. Killing en masse is surely barbaric. Anyone can see that; but it's still being done. It's still "fashionable." Does this mean the guns and bombs themselves are to blame? Or should we rather put the blame on the real villain—man himself? Man simply hasn't evolved far enough yet to do without guns, wars, and killings.

And just as we can't blame the guns for man's barbarism, so we can't blame gold for man's barbarism.

In the Sinai and around the Golan Heights, man has recently been showing his recourse to barbarism as a procedure for settling differences of opinion. Not very advanced, is it? Neither was Vietnam.

Man hasn't perfected any of the techniques for getting along with his fellow man. Or woman. Divorce rates hit new records every year. Crimes of violence are also increasing. Small wonder, if we don't understand our own psychology and know almost nothing of the human brain, that we need a monetary system based on something tangible. In gold we trust. In words we mistrust. And, so far, rightly so.

Man *may* evolve enough in time to escape this dependency. But, in my opinion, it will take a thousand years.

And *until* man has evolved maybe we should be grateful that we have gold, for it will always work; it can help pull us out of a hole. In a world where we send men off to be killed in the name of patriotism or in the name of somebody else's government, in a world where wars are still being fought because of different religious views, in such a world gold doesn't look so terribly barbaric.

It's strange that the politicians who resent the controlling mechanism of gold despairingly refer to gold as a barbaric metal. It is not that they are condemning gold as much as they are pretending that man is elevated above barbarism. He isn't, not really. And until he is, then I must reiterate that, much as we may dislike gold for symbolizing our lack of evolution, gold is going to be around—to scorn us silently when we try to remove it from our monetary system. Gold will be there when, out of chaos and misery, we put our monetary systems back together again. Gold *is*.

We needn't like it, but it is—and I am a realist, not an idealist.

If I may be permitted to offer my vision of the world in 1985 and speculate on what happened to gold, I think the scene would be something like this:

Gold coins had been issued by most countries and were actually in circulation. (They started to appear in the early 1980s even though the money system itself broke down earlier.) Money became so suspect in 1978, and paper currency rates were so erratic, that governments were forced to mint gold coins to reestablish confidence in money. They found out they could legislate what was legal tender but not what is *money*. Central banks began accumulating more gold, buying on the free market, and some bought direct from South Africa. Most major Western nations had built up gold reserves by 50 percent over 1973 levels, some by 100 percent or more.

When the International Monetary Fund was forced, in 1982, to use gold as the de facto cornerstone of the world system, inflation popped like a hot balloon. Prices dropped. This caused some unemployment and business slowed down, but stability had returned and money again had become something you could save without fear that it would lose much of its value while sitting there. There were exchange controls in effect in the United States. That distorted everything, and official U.S. rates and world rates on currencies were vastly different.

All the Harvard professors who grew up bottle-fed on Keynesianism had been repudiated. A new respectability had returned to gold and gold shareholders since the days we can barely remember in the late 40s, 50s, and early 60s when gold was considered something going out of style in this very advanced world.

Actually man wasn't as civilized as he thought. If the premise is valid that man has not evolved enough to push gold aside and substitute fiat money, then clearly one is correct in moving into gold investments. If we are correct in holding that gold's continued ascendency will escalate sharply in the late 1970s and early 1980s, then it is probably too early to be 100 percent in gold. My recommendation is to posture one's portfolio to be at least 50 percent in various forms of gold investments. This percentage should be increased over the next few years to 75 percent to 80 percent by 1978.

A portion of one's investment in gold should be, if possible, in gold bullion itself. This earns no yield and produces storage charges, but it is also free of mine flooding, mine fires, government mine taxes, stock confiscation, and so on. Another portion should be in gold shares, some of which should be high yielding, some low, for diversification. A very small portion should be in gold coins, of common date variety—with minimum premium. Some people may fancy the gold futures market, too, but this is not for the majority.

The breakdown of gold shares should also see diversification between South African shares and North American—mainly Canadian. A 75/25 split is recommended.

As I have pointed out many times, sound-money men tend to love gold because it gives discipline—which gives us stability, a fair measurement, and freedom from inflation. But in fact we should perhaps say we hate gold because man is so unevolved that gold is critically necessary to keep the beast in line.

Do you love the whip because it protects you, or do you hate it because it's still necessary?

Harry D. Schultz

Chapter 1

A Brief History of Gold

"The various necessities of life are easily carried about and hence men agreed to employ in their dealings with each other something which was intrinsically useful and easily applicable to the purposes of life."

—*Aristotle*

Economics textbooks traditionally cover the subject of money by explaining how mankind advanced from bartering salt, wampum, shells, and various other commodities as a means of exchange to trading more sophisticated forms of money. This leads into a discussion of gold, providing an opportunity to serve up all of the hackneyed descriptions and cliches regarding the metal as a "medium of exchange" and a "storehouse of value."

In textbooks, as well as in classroom discussions, the student is swiftly transported to Dullsville.

Yet anyone who makes even a cursory examination of gold's history will find that the metal has had a mystical appeal since the dawn of civilization.

Gold, it is alleged, came even before sex! Some scholars have observed that gold is mentioned in the twelfth verse of the second chapter of Genesis, before the creation of Eve.[1] God didn't make Eve out of Adam's rib until ten verses later.

But aside from this distinction there is no question that gold was one of the first—if not *the* first—metals to be known and used by man.

Croesus, King of Lydia (Western Turkey), is believed to have ordered the first gold coins to be struck in 550 B.C. Like the Pharaohs of ancient Egypt, the Romans, Greeks, Persians, Etruscans, Spaniards, and countless others, he had an insatiable desire for gold.

Man was not long in discovering that, of all the materials used as money over thousands of years, gold is the most nearly perfect medium of exchange and storehouse of value.

Why?

Gold is scarce, beautiful, indestructible, and extremely versatile. A single ounce can be beaten into a sheet covering one hundred square feet and an ounce can be drawn into fifty miles of thin wire. As David Ricardo pointed out, gold is not only an ideal measure of value but also is the *thing* valued. Its two uses, one as a monetary unit and the other as a commercial product, are often described as the "duality" of gold.

Whether it be in jewelry, dentistry, or electronics, gold has no peer in its usefulness. Year after year a relatively unchanging proportion—roughly 50 percent of the world's output of the metal is absorbed by the arts and industry. Jewelry alone accounts for about one-fourth of the total annual use of gold in the non-Communist world.[2] Other industrial uses absorb a comparable percentage; the rest goes into private hoarding or investment, and official coinage use.

We find gold playing an important role throughout the course of history. The California gold rush of 1848-49 not only influenced the monetary history of the world; it also contributed to the metamorphosis of California from a re-

[1]"And the gold of that land is good; there is bdellium and the onyx stone."

[2]Most of the gold used in jewelry is 14-karat in a scale of quality (purity) ranging from 1 to 24. Thus, 14-karat gold contains 14/24 gold or 58.35 percent.

mote and unsettled region to a bustling area. In the process, it caused San Francisco to be built in little more than a year instead of decades.

In Czarist Russia, too, gold acquired an important status in the economy that, incidentally, persists to this day, despite the fact that capitalist and communist worlds may be poles apart politically and philosophically.

Lenin once predicted that gold was only a temporary necessity and that after socialism triumphed over capitalism the yellow metal would be relegated to such functions as covering the floors of public lavatories. He turned out to be a poor prophet, just as misguided as the Keynesian economists who attacked gold as an outmoded metal destined for obscurity.

Not long after gold was discovered in Russia, a short-lived gold rush took place in Australia. Once again it produced a frenzied search by shoemakers, bakers, butchers, clerks, and all sorts of ordinary folk who joined in the hunt for quick riches.

The South African gold rush came years later. It proved to be a bit different from previous gold discoveries. In the vanguard of this boom were legendary figures like Cecil Rhodes, who already had struck it rich elsewhere; with some of his contemporaries, Rhodes had acquired a fortune in the diamond fields of South Africa. This, in turn, provided the capital needed to undertake the costly development of the South African gold mines, mines that have since become by far the world's largest single supplier of the yellow metal.

Gold in Modern Times

Famed economist John Maynard Keynes once dismissed gold as a "barbaric relic."

Actually, the worldwide acceptance and consistent demand for gold would justify a title of the "noblest of metals" in the twentieth century.

Until the outbreak of World War I, the monetary systems of the Western nations were closely linked to gold, and the gold standard performed relatively well.[3] But the war-induced outflow of gold from Europe so weakened the backing of European currencies that all of the major nations of the world, with the exception of the United States and Switzerland, abandoned the gold standard and suspended specie payments.

Leading economists of that era, including Lord Keynes of England and Gustav Cassel of Sweden, argued that "gold has failed, both as a means of payment and a standard of value." Subsequent American monetary legislation came to embrace this theory, based on the false premise that the gold standard was responsible for most of the economic ills from which the world suffered.

[3]When economists refer to the "gold standard," they usually allude to not one, but several types of monetary systems linked to gold.

The gold bullion standard, for instance, obliged the central bank of a country to sell gold in stipulated quantities at a fixed price to meet international obligations.

The gold exchange standard called for the inclusion of foreign exchange as part of the legal reserve of the central banks on this standard.

The "limping gold standard," on the other hand, involved a percentage coverage of currency by a gold reserve backing.

This theory of blaming gold for man's blunders has been effectively demolished by many competent authorities, including Madden and Nadler[4] who, though by no means advocates of gold in all of its aspects, concluded that as a standard of value, as a basis for currency and credits, and as a medium of settling balances among economic political units, "experience has shown that no other metal can better perform this function and the complete abandonment of gold would lead to currency chaos."

Today, despite the U.S. Treasury's long-standing attempts to downgrade gold as a monetary metal, the U.S. still holds a fourth of the world's total gold stock, as against more than two-thirds in 1949. Germany is the second largest Free World holder of gold and has about 10 percent; France has 9 percent; Switzerland and Italy about 7 percent each; and the United Kingdom about 2 percent.

United States Gold Policy

The United States, over the course of its history, has had shifting monetary standards. They have shifted so much that a look at the past is worthwhile, just as a matter of curiosity.

From 1862 to 1879 the nation was essentially on an irredeemable paper standard. U.S. notes (which came to be known as greenbacks) were issued in amounts far exceeding the nation's monetary requirements for that period. Since they weren't redeemable in gold on demand, naturally it was not long before they drove gold out of circulation—people kept their gold and did not spend or circulate it.

This was just one of many historical illustrations of Gresham's Law, which holds that bad money drives good money out of circulation.

Under the Coinage Act of 1792, a bimetallic standard had been established in which the U.S. dollar was defined as 24.75 grains of fine gold or 371.25 grains of fine silver—a 15-to-1 ratio. An ounce of gold, therefore, was worth $19.42. But this ratio overvalued silver at the mint. Silver proceeded to drive gold out of circulation—only silver circulated for much of the period from 1792 to 1834, though the country ostensibly was on a bimetallic standard for that period.

An 1834 law reduced the pure gold weight of the dollar from 24.75 grains to 23.2 grains, and the price of an ounce of gold was raised to $20.67. Now, gold was overvalued; this helped drive silver out of circulation.

In order to avoid an outflow of gold during the Civil War and thereafter, the nation went off the metallic standard in the 1862-1879 period. Paper currency no longer was redeemable in gold or silver.

In February 1873 the infamous "Crime of '73," as the antisilver legislation was called by the silver interests, was enacted; it dropped the silver dollar from the list of standard coins to be minted. The outcry led to the passage of the Bland-Allison Act in 1878, which required the Secretary of the Treasury to purchase silver to be added to the national monetary stock.

[4]*International Money Markets*, Prentice-Hall, 1936.

VOL. XCVII. No. 3 NEW YORK, WEDNESDAY, OCTOBER 30, 1929 88 PAGES

WALL ST. LAYS AN EGG

Going Dumb Is Deadly to Hostess In Her Serious Dance Hall Profesh

A hostess at Roseland has her problems. The paid steppers consider their work a definite profession calling for specialized technique and high-power salesmanship.

"You see, you gotta sell your personality," said one. "Each one of we girls has our own clientele to cater to. It's just like selling dresses in a store—you have to know what to sell each particular customer.

"Some want to dance, some want to kid, some want to get soupy, and others are just 'misunderstood husbands'."

Girls applying for hostess jobs at Roseland must be 21 or older. They must work five nights a week. They are strictly on their own, no salary going with the job and the house collecting 10 cents on every 35 cent ticket. To keep her job, a girl must turn in at least 100 tickets a week

Hunk on Winchell

When the Walter Winchells moved into 204 West 55th street, late last week, June, that's Mrs. Winchell, selected a special room as Walter's exclusive sleep den for his late hour nights. She shusshed the Winchell kidlets when her husband dove in at his usual eight o'clock the first morning.

At noon, Walter's midnight, his sound proof room was penetra d by so many high C's he awoke with but four hours of dreams and a grouch. Investigated at once, after having signed the lease of course.

Right next door, on the same floor, is the studio of the noted

DROP IN STOCKS ROPES SHOWMEN

Many Weep and Call Off Christmas Orders — Legit Shows Hit

MERGERS HALTED

The most dramatic event in the financial history of America is the collapse of the New York Stock Market. The stage was Wall Street, but the onlookers covered the country. Estimates are that 22,000,000 people were in the market at the time.

Tragedy, despair and ruination spell the story of countless thou-

Kidding Kissers in Talkers Burns Up Fans of Screen's Best Lovers

Talker Crashes Olympus

Paris, Oct. 29.

Fox "Follies" and the Fox Movietone newsreel are running this week in Athens, Greece, the first sound pictures heard in the birthplace of world culture, and in all Greece, for that matter.

Several weeks ago, Variety's Cairo correspondent cabled that a cinema had been wired in Alexandria, Cleopatra's home town.

Only Sodom and Gomorrah remain to be heard from.

Boys who used to whistle and girls who used to giggle when love scenes were flashed on the screen are in action again. A couple of years ago they began to take the love stuff seriously and desisted but the talkers are reviving the ha ha for film osculators.

Heavy loving lovers of silent picture days accustomed to charming audiences into spasms of silent ecstasy when kissing the leading lady are getting the bird instead of the heartbeat. The sound accompaniment is making it tough.

Such a picture romancer as John Gilbert is getting laughs in place of the sighs of other days, and the flaps who still think he's grand are getting sore. One little flap had to be quieted by an usher when making a commotion during a Gilbert picture at the Capitol, New York

The famous Variety *headline of October 30, 1929, proclaiming the Wall Street Crash*

The U.S. Bullion Depository, Ft. Knox, Kentucky

Gold Bars in the U.S. Bullion Depository, Ft. Knox, Kentucky

A True Gold Standard

In 1879 the nation returned to the redemption of greenbacks in gold. But a true gold standard wasn't instituted until the defeat of the silver interests, led by William Jennings Bryan, and the passage of the Gold Standard Act of 1900. This provided that the dollar, consisting of 28.5 grains of gold nine-tenths fine (23.22 grains fine gold), should be the standard unit of value. The nation was finally on a single gold standard with a statutory price of $20.67 an ounce.

Except for the 1917-1919 period when the disruptive effects of World War I compelled an embargo on gold exports (the embargo was lifted in 1919), the Government's buying and selling price for gold remained $20.67.

Roosevelt Devalues

The Depression and the Gold Purchase Plan of 1933 brought about historic changes. On October 22, 1933, President Franklin D. Roosevelt said, in a historic radio address:

> Because of the conditions in this country and because of events beyond our control in other parts of the world, it becomes increasingly important to develop and apply further measures which may be necessary from time to time to control the gold value of our own dollar at home.
>
> Our dollar is now altogether too greatly influenced by the accidents of international trade, by the internal policies of other nations and by political disturance in other continents.
>
> Therefore, the United States must take firmly in its own hands the control of the gold value of our dollar. This is necessary in order to prevent dollar disturbances from swinging us away from our ultimate goal—namely the continued recovery of our commodity prices.
>
> Therefore, under the clearly defined authority of existing law, I am authorizing the Reconstruction Finance Corporation to buy gold newly mined in the United States at prices to be determined from time to time after consultation with the Secretary of the Treasury and the President. Whenever necessary to the end in view, we shall also buy and sell gold in the world market.

Just three days later the official price of newly mined gold for that day was fixed at $31.36 an ounce. This was slightly above the world price.

Arthur M. Schlesinger, Jr., in *The Age of Roosevelt*, relates how Roosevelt, aided by Secretary of the Treasury Henry Morgenthau and Chairman Jesse Jones of the RFC, would fix the price of gold each morning while Roosevelt ate his eggs and drank his coffee. Not infrequently they joked about "lucky numbers" and what the day's price would be.

Unfortunately the Rooseveltian blueprint didn't work. It soon dawned on FDR that his moves weren't having the tonic effect on the economy that he had hoped for. So, under the authority of the Gold Reserve Act of 1934, Roosevelt by proclamation on January 31, 1934, fixed the weight of the gold dollar at 13.71 grains of fine gold and the dollar price of gold at $35 an ounce instead of the old price of $20.67. It was announced to the world that the United States would stand ready to buy and sell gold at $35 an ounce.

The Gold Reserve Act of 1934, unfortunately, not only raised the price of gold: it also prohibited the private ownership of gold. Only Americans using gold for industrial purposes were given licenses to obtain the metal, based on their proven needs.

The $35-per-ounce price prevailed from 1934 until 1971, when the official price was raised to $38 an ounce by President Nixon. Only fourteen months later Nixon again devalued the dollar, in February 1973, when the price of gold was raised to an official rate of $42.22 an ounce.

Implicit in the action was an admission that the U.S. Treasury's inflexible policy toward gold pricing had been a failure.

As Milton Gilbert, Economic Advisor of the Bank for International Settlements, has stated so eloquently: "The long resistance of the United States to a justified adjustment of its position by an adequate increase in the official dollar value of gold must certainly be the greatest self-imposed infliction in monetary history."

Chapter 2

Gold, Oil and Inflation

"The prices vary much on the road and the eternal confusion with the good and bad money and its different value is enough to weary a bank clerk."

—*Diary of a traveler from London to Vienna in 1829*

Any currency, regardless of the importance or strength of the country involved, needs some backing. Thousands of years of history have proven that as soon as a currency departs from its gold backing, that currency becomes a commodity—and commodities fluctuate according to supply and demand.

History also demonstrates time and again that an unbacked paper currency can only go down in value. That is what has happened in the United States. Since the Thirties, the money supply has tended to increase faster than the output of goods and services, creating built-in inflation that has now reached double-digit proportions.

Monetary expert Nicholas L. Deak, President of Deak & Perera Corporation, the oldest and largest foreign money exchange firm in the Western hemisphere, has defined inflation in terms that anyone can understand.

"Inflation occurs when a society is trying to do more than its resources permit. There is no such thing as controlled inflation. Rather you see creeping, walking, or runaway inflation. It builds up speed slowly, then goes faster and faster."

With the U.S. Government constantly running budgetary deficits—that is, spending more than it takes in—more purchasing power has been put into the hands of the public than is withdrawn via taxation. Those continuing increases in the money supply have not been offset by gains in productivity and so we have people paying higher prices for a limited amount of goods offered for sale. Labor demands more wages to offset the higher prices, corporations raise prices to offset their higher costs, then labor comes back demanding still more, and the process goes on and on.

The results have been catastrophic. According to the First National City Bank, "never since the end of World War II has the purchasing power of money dwindled so rapidly as in the first half of 1974." For most countries, including the United States, two-digit rates of inflation now are the rule rather than the exception.

It is, of course, possible to blame part of today's runaway inflation on the soaring oil prices that stunned the Free World in the wake of the 1973 Arab oil embargo. But the energy crisis only exacerbated an already deteriorating situation. It was already evident that politicians were caving in before the pressures of inflation in one country after another.

The Tory Government in Great Britain—not the Laborites—increased the money supply at rates that touched 33 percent yearly. And in the United States both Democrats and Republicans knuckled under to inflationary pressures.

Democrat Lyndon Johnson made little or no effort to finance the Vietnam War with a tax increase, contending that America could afford both "guns and butter." By ruling out a politically unpalatable tax boost to finance the most unpopular war in U.S. history, he laid the foundation for a massive inflationary groundswell later on.

Republican Richard Nixon did little better. He, too, failed to come to grips with the economic problems facing the nation, much like his predecessor.

Although he brought the Vietnam conflict to a conclusion, Nixon amassed huge budgetary deficits and, worse yet, he permitted inept management of wage and price controls by his Treasury Secretary George Schultz that only added more fuel to the inflationary fires.

After Watergate washed Nixon out of office, his successor, Gerald Ford, displayed zeal for combating inflation by coming up with a new slogan—"WIN," for Whip Inflation Now. That about sums up his program.

Is it any wonder, in the light of this sorry record, that the United States has been battered by a collapsing stock market, the highest interest rates in the nation's history (because the Federal Reserve System is trying to do what the politicians won't do), and the most severe erosion of public confidence since the Great Depression of the Thirties?

While this sorry economic record was being compiled, the U.S. Treasury's distaste for gold remained constant. In 1961 President Kennedy put an end to the right of U.S. citizens to own gold bullion overseas. In 1965 President Johnson eliminated the 25 percent gold backing for Federal Reserve deposits. And President Nixon twice devalued the dollar within the space of fourteen months.

The Impact of Oil

As the world faces inflation, recession, and unemployment, it also must deal with the incredibly complex problems created by the quadrupling of oil prices.

The monopoly power of a tiny handful of oil-producing nations gives them the ability to cripple the economies of the Free World. The Organization of Petroleum Exporting Countries (OPEC) nations are accumulating surplus funds at the rate of $70 to $80 billion a year, while the rest of the world is running deficits of the same massive proportions.

The World Bank estimates that the cumulative monetary reserves of the oil-producing countries—and the debts of the rest of the world—will rise to $650 billion by 1980. By 1990 the total will reach $1.2 trillion dollars.

This represents the greatest transfer of wealth in the world's history and it is questionable whether the banking system can meet the challenge presented by the oil-price explosion. Already the financial markets are facing difficulties in "recycling" the huge petrodollar revenues being accumulated by Arab and other oil-exporting countries back to consuming nations in the form of loans and investments.

Of course, even if recycling works in terms of shifting monetary balances, it offers no solution to controlling the inflation that the huge oil-price hikes have created. Already standards of living are falling in Western Europe; Italy is virtually bankrupt, Great Britain is not far behind, and France and Spain may be only a year or two from economic catastrophe.

The SDRs Created

One of the unique inventions of the monetary experts has been the so-called "Special Drawing Rights" for gold (or SDRs) that were unveiled by International Monetary Fund officials in 1967. They were supposed to fulfill a role as a "new international reserve asset."

The SDRs were simply bookkeeping entries, created by the stroke of a pen. Each country was assigned a quota, based on its contribution to the IMF. The values the SDRs possessed resulted solely from the fact that the central bank members of the International Monetary Fund pledged to accept them in settlement of their balances with each other.

Originally SDRs were defined in gold; but, inasmuch as they were not actually redeemable in gold, they came to be known as "paper gold."

The optimism expressed by the creators of the new paper gold soon wore thin. In 1974, with the free market price of gold far above the official price, the SDRs were redefined. Instead of being expressed in gold, they were expressed in some sixteen paper currencies, most of which had been depreciating in value for years. Small wonder then that the governors of the International Monetary Fund have difficulty in agreeing to any basis for allocating them among member nations.

"Floating" Rates

Another gimmicky cosmetic solution to the deep-rooted monetary problems of the world has been "floating rates." The argument advanced in their behalf is that countries with fixed exchange rates cannot effectively determine their own money supplies and therefore their own price levels. It would be better, so the theory went, for currencies to "float" against each other without any fixed ratios of exchange.

What this system actually does is increase the cost of dealing internationally by increasing the cost of currency contracts. The reason is simple: it has become absolutely essential for international traders of goods or commodities to insure the value of the funds they will receive.

In fact, floating rates not only discourage world trade by adding to costs, but they also generate additional problems. For example, they invite overspeculation in international currencies, as attested by the plight of some of the best-known names in the world banking community. And they also encourage governments to manipulate their currencies to create favorable balance of trade positions without going through the more painful process of making fundamental internal changes in their economies.

Otto E. Roethenmund, vice-chairman of the Foreign Commerce Bank of Switzerland, warns that floating rate problems may be with us for some time. He declares that "floating rates were supposed to be a temporary measure, but this is not so . . . we will have floating rates until there are major changes in the currencies of the western world."

Bankers' Problems

The effect of international currency instability on the banking institutions of the world is not to be overlooked. As *The New York Times* reported on

October 7, 1974, "from the provincial backwater of Orangeburg, S.C., to the glittering financial centers of New York, London, Zurich, and Vienna, banks are no longer the rocks of Gibraltar they once were. Hardly a week goes by without some new banking problems surfacing."

The problems of the now-defunct Franklin National Bank (N.Y.) were once described as untypical of banks as a whole. But it has since become evident that Franklin was just the tip of an iceberg—the troubles of the banks are clearly worldwide. Institution after institution—including some of the world's most prestigious names—have "discovered" irregularities or unusual losses of one kind or another. No one could be unduly startled by the goings on at the Beverly Hills Bancorporation or the United States National Bank of San Diego. But the list of institutions with unexpected problems include the Chase Manhattan Bank, Lloyds Bank of London, and the Union Bank of Switzerland. Early in 1974 Bankhaus Herstatt in Cologne, West Germany, went bankrupt after suffering heavy foreign exchange trading losses, and during subsequent months several smaller European banks were closed.

There is little doubt that the obsession with performance that bankers have succumbed to in recent years has now come home to roost. And while some contend that the Federal Reserve couldn't possibly permit one of the largest U.S. banks to go under, this is scarcely true of hundreds of smaller banks that could become mired in financial difficulty. The overall effect of any sizable rescue operation would doubtless be highly inflationary.

In retrospect there is little doubt that the "gunslingers" of 1974 were the one-bank holding companies that were in an analogous position to the youthful stockbrokers who were being hailed as financial geniuses just a few years ago.

It must be remembered that the funds that represent the capital of the banks have their base in fiduciary sources. Yet these acquisitive enterprises have been permitted to enter fields that bear no relation to fiduciary responsibilities—for instance, they have engaged in speculative real estate syndications, high-risk foreign exchange gambits, and provided the wherewithal for purely speculative corporate take-overs.

Ironically, the adventurous commercial bankers, hell-bent on achieving a "multiple" for their stocks, have even been permitted to siphon off funds from the savings banks and other savings institutions that normally provide the bulk of the funds for home mortgages, by offering sliding-rate capital notes with rates linked to the U.S. Treasury bill rate.

Inflation's Manifestations

There is no difficulty in singling out numerous manifestations of inflation in the American economy in 1974.

Example: The Lincoln Savings Institution, a major New York City savings bank, wrote mortgage holders offering to accept $12,831 in full payment of a $15,000 mortgage.

We have a long distance to go before we reach the stage where inflation is unbearable, where prices change not monthly or weekly, but day by day and even hour by hour.

But the handwriting is on the wall. We now have escalation clauses in virtually all major labor contracts because unions won't sign up without some assurance that their pay raises won't be wiped out by rising prices.

Because of inflation there is growing interest in *indexation*—compensation in just about every financial transaction for the changing rate of inflation. It has already met with some success in Austria, which has adopted an interest bonus scheme calling for extra payments to a lender, equal to the rate of inflation. Brazil seems to have had some success with indexation. More will be said about this inflation-compensation system later on.

Effective antiinflation measures are still possible even at this late date. But unfortunately they must be relegated to the realm of wishful thinking because they involve politically unacceptable risks.

Any program to check inflation inevitably means increased unemployment and politically unpalatable tax hikes. Yet, even with inflation in America running over 10 percent annually, members of Congress are calling for measures to counter deepening unemployment and are demanding tax reductions—not increases.

In other words, we cannot count on politicians to act responsibly. Their first responsibility is to get reelected, even if it means more inflation.

Americans, naturally, aren't as aware of inflation's perils as are most Europeans. Those Americans who examine photographs of Germans burning worthless marks issued during the Weimar Republic can only feel a sense of curiosity. They are confident it couldn't happen here.

Germans, however, feel quite differently. Some can still remember the days in 1923 when a newspaper cost 200 marks, with the price only three months later jumping to 2,000 marks.

At one stage a half-liter of milk went for 250,000 marks; lodgings 400,000 marks; dinner 1.8 million marks.

In the summer of 1922, a dollar was worth 575 marks. Only a year later it was worth 1.65 billion marks!

Superinflation had a traumatic effect on Germany and helped bring Hitler and Nazism to power, with tragic consequences for the Germans as well as the rest of the world.

The Germans haven't forgotten. That is one reason why West Germany has controlled inflation better than any other leading industrial power in the Western World.

The Role of Gold

But what if the United States should get into an uncontrollable inflation?

We would then have to prepare for a new currency whose monetary value could be defended. This could be accomplished only with a gold-linked dollar, in combination with a steep rise in the official gold price, and a reopening of the gold window for central banks.

Gold, of course, would not be the only answer to such a dilemma.

However, it is the only constituent of the monetary reserves in the Western World with the ability to grow by virtue of price increases.

Proof of this was furnished by an agreement entered into by West Germany and Italy in 1974. For purposes of the loan agreement, Italy's gold was valued at the free market price averaged over an eight-week period, or approximately $120 an ounce. This collateralized value was far in excess of the official U.S. Treasury price of $42.22 per ounce.

Italy, in other words, issued gold-backed promissory notes to obtain a $2-billion credit from the Germans. And the Germans accepted (as collateral) gold valued near the free market price rather than the artificial "official" U.S. Treasury price. For those hoping to minimize gold's importance in international monetary transactions, the German-Italian deal was nothing less than a death blow.

More accommodations of this type may be expected on the international scene, inasmuch as other countries are facing severe financial strain because of soaring oil bills.

One solution—perhaps the only solution at the moment—may be gold-collateralized transactions.

However, a word of caution is in order. The Italian-German agreement, concluded in an atmosphere of crisis, assisted Italy by providing funds needed for its oil purchases. But perhaps historians will conclude that it might have been better for Italy to default on her oil obligations and thus bring pressure on the Arabs to ease their stranglehold on the Free World's economy.

While the use of gold at market-related prices obviously can serve to increase the reserves of a country entering into such an arrangement, the benefits are not all one-sided. As reserves are increased without benefit of a monetary system with controls on the use of such increased reserves, a new "engine" of inflation is created. (The financial press hasn't realized this yet.)

This means that even though such agreements do provide an easy way of meeting an emergency situation, such as paying an oil bill, no positive long-term economic benefits are gained unless the agreements are also accompanied by a revised monetary system that acts to control the uses of the newly created reserves.

The solution to the problem, therefore, must start at the root cause. A monetary system must be devised that is sound, viable, and disciplinary—with the help of gold. Not until such a system is devised, and its responsibilities accepted by the international monetary authorities, will the winds of economic trauma and disorder subside.

Chapter 3

American Ownership of Bullion

"People fight the gold standard because they want to substitute national autarchy for free trade, war for peace, totalitarian government omnipotence for liberty."

—*Ludwig von Mises*

On August 14, 1974, President Gerald Ford signed a bill allowing American citizens to buy and sell gold on and after December 31, 1974. Many believed that this event would not be allowed to take place. In fact, for some inexplicable reason, Dr. Arthur F. Burns, chairman of the Federal Reserve Board, waited until just prior to legalization to trumpet that, "it is my duty to point out that prompt removal of present restrictions on private trading in gold could complicate a financial situation that is already beset by strains and stresses."

Nevertheless, the most momentous event to affect gold in nearly half a century did take place, and the result was a disappointment for those who had expected fireworks.

The initial American reaction was one of apathy, and, as one London dealer described it, "the whole thing came off like a slightly damp squib," a damp squib being the British equivalent of a wet firecracker.

What happened?

The explanation seems clear. In anticipation of heavy demand from American gold buyers, the price of gold was pushed up to an all-time high of $197.50 an ounce on December 30, just a day before Americans could legally buy the yellow metal. When American demand failed to meet expectations, investors and speculators rushed to bail out and the result was a sharp slide in the price of bullion, which plummeted $11.50 an ounce in a single day.

The lack of American interest in gold in the period immediately following legalization can, in our judgment, be attributed to four factors:

1. The anticipatory price rise in gold just prior to "G-day" when American ownership was legalized.

2. The well-publicized Treasury announcement that it would auction two million ounces of gold, worth about $380 million at prevailing market prices, less than a week after the legalization date.

3. The disclosure that major banking institutions, such as the Bank of America, First National City Bank, and Chase Manhattan Bank, had decided not to offer gold at retail.

4. Warnings from a host of official and unofficial sources concerning gold's price volatility, high acquisition cost, and other risk factors.

A Healthy Beginning?

We are convinced that the failure of a buying panic to develop in the initial phases of gold trading in the United States is healthy for the future of gold ownership by Americans. Nothing dampens the ardor of a speculator or investor more than buying at the top of a price rise and then being obliged to sit with paper losses for many months. Fortunately, Americans proved to be more canny traders than some Europeans had expected and didn't rush in to buy gold at the highest prices on record.

Incidentally, many observers overlooked the fact that impressive activity took place in the futures market, where the volume of gold futures contracts was the highest for any opening day of a new contract on U.S. exchanges.

We have long contended that it would be most unfortunate if the American merchandising miracle that led to mass consumption of soap, cigarettes, and deodorants became so pervasive that it branched out to include gold. Slick promotion may be fine for household necessities, but when it's aimed at encouraging gold ownership it would be self-defeating. It could only result in bringing many people into the gold market who do not understand the nature of the metal—ultimately this would have negative consequences.

This was certainly true in the case of silver, which was excessively promoted to an audience which didn't comprehend its speculative nature. For many uninformed buyers, silver's glitter proved just a mirage.

United States vs. Japan

Some skeptics have noted that in Japan a short-run buying spurt soon faded after gold ownership was legalized—and they have predicted that a similar pattern will emerge in this country. However, in evaluating the ephemeral interest of Japanese investors in gold, it must be emphasized that, historically, the Japanese have displayed a stronger interest in *platinum* than have Westerners. Also, it seems clear that the failure of Japanese suppliers to "make a market" in gold bars made it difficult for investors to resell their gold, and, naturally, this helped to dampen the demand for gold. The special features of the Japanese market (as well as the experience of other countries) must be taken into consideration before assuming that America's appetite for gold will simply be a carbon copy of Japan's.

Traditionally it is true that Americans are not congenital gold hoarders—they apparently turn to gold only when other values appear to be threatened with collapse. In contrast, the French, after centuries of war, revolution, and currency devaluations, have learned to love gold. It is estimated that the French account for nearly half of the privately held hoard of gold in Europe. It would take a bold forecaster to predict that Americans will acquire the same passion for the metal that has become traditional in other parts of the world.

But who is to say that American psychology is immune to change, particularly if a great number of people become convinced that the United States is on the brink of a cataclysmic monetary and economic crisis?

This is the enigma faced by forecasters seeking to assess future demand for gold by Americans.

Among the negativists on American ownership of gold is Thomas W. Wolfe, director of the Domestic Gold and Silver Operations of the Treasury Department. Wolfe, who has declared that gold will fail to captivate most Americans, has pointed out that half of all the gold produced since the beginning of time is now in official bank reserves.

He has declared, ominously, that "there's about 1.1 billion ounces, literally thousands of tons of the stuff, sitting in government vaults like a monster, an atom bomb that can be unleashed at any time. All you've got to do is back up a thousand trucks and it's ready in a nice clean, airy hall. No other commodity operates under such a deterrent. So gold is valuable only as long as it's not put on the market."

But even while offering these negative portents, the same Treasury official has predicted that "the primary trading center for the world gold market will ultimately shift to the United States and this change may indeed be quicker than some would expect."

Obviously, the American situation is quite different from that of major European countries where the purchase of gold has been compared to stopping off at the butcher's shop to pick up a pound of meat for the dinner table. America will not become another France or India overnight.

It also must be kept in mind that, even while the decades-old prohibition against gold ownership remained in effect, some Americans found ways to satisfy their desire for the yellow metal. For instance, they could buy gold coins, jewelry, the shares of gold mining companies and other forms of gold-related investments.

They also dabbled illegally in gold purchases. No one can offer more than an educated guess, but some authorities have estimated the value of gold held by Americans in banks outside the United States at anywhere from $5 billion to $10 billion.

It would be well to note that there have been no parallel situations in the United States for over forty years and therefore any predictions about the attitude toward gold ownership over the long run must be regarded as opinion rather than fact. However; for our part, we don't doubt in the least that the lifting of restrictions on the right of an American citizen to own gold is an extremely positive development for this commodity that commands confidence in every corner of the globe.

We would not be surprised, however, if investors and speculators took a "wait and see" attitude toward gold for many months, but we completely disagree with those who have dismissed gold fever as just a "24-hour virus." The record has yet to be written, and we regard legalization in America as unquestionably bullish for gold on a *long-term basis*.

It should be borne in mind that gold is not so much an investment as a speculation. It cannot be considered a pure investment because it produces no current interest return for the holder. Actually, the lack of any return on the investment, plus storage and insurance charges and possibly a premium for purchases in small quantities, results in a negative return for the holder. A rough rule of thumb is to add somewhere around 15 percent to 20 percent per year to the nominal purchase price to reflect these costs.

Then, too, remember that the gold market—unlike the securities market—is almost totally unregulated. The Securities & Exchange Commission, though it emphasizes that all sales of gold aren't necessarily exempt from securities laws, and that it will be closely scrutinizing future developments, has also indicated that it doesn't consider gold sales by banks and brokerage firms as being subject to regulation in the same manner as securities.

GOLD PRODUCTION
(In millions of troy ounces)

	1970	*1971*	*1972*	*1973*	*1974 est.*
South Africa	32.2	31.4	29.3	27.4	27.4
U.S.S.R.	6.5	6.7	6.9	7.1	7.3
Canada	2.4	2.3	2.0	1.9	2.0
U.S.	1.7	1.5	1.5	1.5	1.5
Other	4.7	4.6	5.2	5.5	5.4
Total:	47.5	46.5	44.9	43.4	43.6

GOLD HOLDINGS
(In millions of U.S. dollars)

	1970	*1971*	*1972*	*1973*
United States	$11,072	$11,081	$10,487	$11,652
Germany	3,980	4,426	4,459	4,966
France	3,532	3,825	3,826	4,261
Switzerland	2,731	2,158	3,158	3,513
Italy	2,887	3,131	3,130	3,483
Netherlands	1,787	2,072	2,059	2,294
Belgium	1,470	1,676	1,638	1,781
Portugal	902	1,000	1,021	1,163
Canada	791	862	834	927
Japan	532	738	801	891
United Kingdom	1,349	842	800	886
Austria	707	791	792	881
South Africa	666	445	681	802
Sweden	200	217	217	244
Denmark	65	69	69	77
Norway	26	36	37	41

Source: International Monetary Fund.

LONDON GOLD PRICES/JANUARY 1972 TO AUGUST 1974

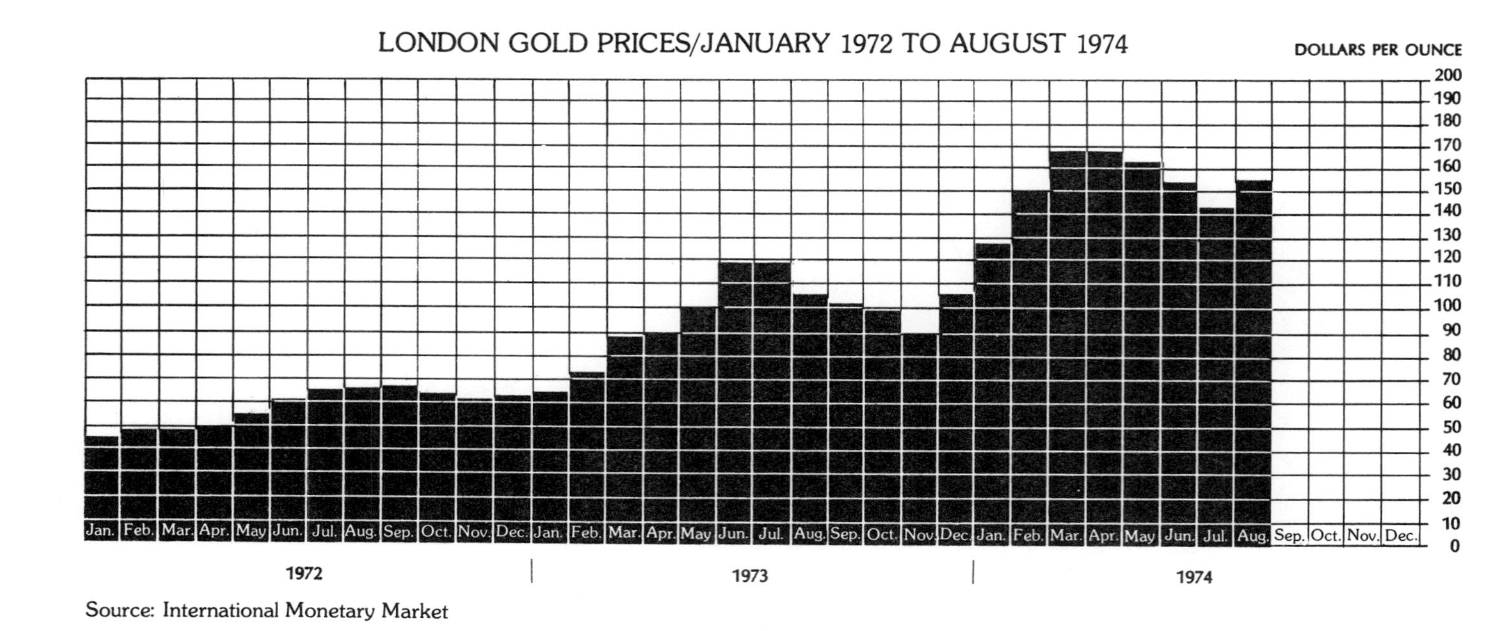

Source: International Monetary Market

This means that the prospective gold buyer must follow the rule of *caveat emptor* and be sure that he is dealing with a reputable firm, preferably one that has been in business for a long time. (If you aren't sure, check with your bank or Better Business Bureau.) Avoid doing business with a hastily organized one-man "mint" which may resort to "bait" advertising in leading newspapers and magazines. It is unfortunate but true that sometimes the most unscrupulous do the most advertising.

These precautions are essential for anyone who wishes to minimize risks in buying either bullion or coins where there are risks of counterfeiting, varying price quotations for comparable items, and potential problems in finding a buyer when you may be ready to sell.

Where to Buy Gold

Banks, brokerage firms, coin and bullion dealers, private mints, department stores, jewelers, and many other outlets have been ready to supply gold since the ban on American ownership was lifted.

Some of the offerings, of course, may be considered as in the novelty category. For instance, during the 1974 Christmas period, a number of New York City jewelers reportedly did a big business in selling miniature replicas of gold bars, "just like the ones in Fort Knox." A one-ounce bar, an inch long and about a half-inch thick, sold for $360, or about double the then-prevailing price for bullion on the world market.

Gold bars may well be a popular gift item for the man or woman who "has everything." But since Cartier and other retailers sold the items as jewelry, anyone seeking to resell these expensive gimmicks will be in for a king-sized headache.

Brokerage Houses

The serious gold buyer would do well to consider the services of leading stock brokerage firms which have entered the gold business. For instance, Merrill Lynch, Pierce, Fenner & Smith; Bache & Co.; and Paine Webber, Jackson & Curtis, among others, have established programs for dealing in gold. A smaller New York Stock Exchange firm, Samuel Weiss & Co., was one of the first to announce plans to handle gold.

Vilas & Hickey, a New York City-based New York Stock Exchange member firm, of which one of the authors of this book is a general partner, has facilities to handle orders for either gold bullion or gold coins at competitive prices, adjusted hourly in accord with world markets.

Merrill Lynch has joined with Samuel Montagu & Co. Ltd. and Handy & Harman in forming a jointly owned precious metals company (Merrill, Montagu, Handy & Harman) which acts as a gold dealer for U.S. institutions, corporations and citizens; it will enable anyone to purchase and sell gold through the worldwide offices of Merrill Lynch. Wholesale customers such as banks, department stores, and jewelers are offered gold bars in these sizes: ½, 1, 5, 10, 50, 100 and 400 ounces, as well as bars of one kilo (32 troy ounces). Individuals will be able to purchase bars in multiples of five ounces.

Typically, brokerage customers will pay the London Market's opening price for the morning following the buyer's order. (The opening price, which is known as the London Bullion Fix, is issued at 5:30 A.M. New York time.) Gold may be sold on the same terms that it is bought—the London market price for the following morning, less the charge. (The once-a-day "fix" plan for buying and selling will undoubtedly give way to a more flexible plan as experience is gained in buying and selling gold.)

In order to be competitive, brokers must offer gold that bears a well-known assay mark. They should offer to deliver gold to the customer's bank for safekeeping or else provide the machinery whereby gold can be stored and the buyer provided with a document similar to a warehouse receipt.

Incidentally, the New York Stock Exchange has taken steps to facilitate gold and silver dealings by its member firms. The Exchange has proposed a rule to permit spot bullion dealings on credit by the public without charge against a broker's capital so long as the transaction meets certain criteria, including an acceptable custodian and appropriate insurance. Credit would be sanctioned up to 75 percent margin (up to 90 percent if hedged by bullion futures). Payment in five business days would be classified as a cash transaction.

Banks and Gold

The banking industry appears to have taken a generally cautious attitude toward dealing in gold. A number of the largest banks (including Bank of America, First National City Bank of N.Y., Chase Manhattan Bank, Manufacturers Hanover Trust Co. of N.Y., Chemical Bank of N.Y., Mellon Bank of Pittsburgh, and Cleveland Trust Co.) have decided to stand on the sidelines for the time being and shun retail dealings in gold. Mellon Bank, for instance, stated that it feared the promotion of gold sales "might induce speculation by individuals who don't have a clear understanding of the risks involved."

On the other hand, Bankers Trust Co., Marine Midland Bank, Sterling National Bank, and Republic National Bank, all of New York, are handling gold transactions. Marine Midland, for instance, will sell a minimum of five ounces of gold, and its fees include a 5-percent sales charge, a fabrication fee, and delivery and storage charges.

Sterling National Bank has available at its teller windows one-half-ounce, one-ounce and five-ounce bars, at the prevailing price at the time of purchase, plus a fee estimated at 6 percent to 7 percent.

Republic National Bank, the largest U.S. gold bullion dealer selling to licensed industrial users, also is serving individuals interested in owning gold because, as a spokesman has said, "It is just an extension of what we have been doing."

One handicap banks face in the early stages of American ownership is the lack of regulations governing their handling of gold. Then, too, today's banking personnel lack the training to handle gold transactions efficiently. And, naturally, the cautious nature of bankers impels them to wait and see if the gold bullion market reaches the $3.5-billion level some authorities have estimated for 1975, or whether it will be considerably below this level.

As Charles C. Smith, a vice-president of the Bankers Trust Company of N.Y., has commented, "It's a little like the end of Prohibition, but probably a more complete change."

Bankers, of course, are worried that some unsophisticated customers might hold banks responsible if they were unable to sell gold at the price they originally paid for it. The banks are also concerned over the lack of qualified personnel, and of course they can hardly be expected to welcome large-scale withdrawal of funds from checking and time accounts during a period of credit stringency.

Still, banks must be regarded as a logical marketing force in gold transactions once they overcome their initial hesitancy. They can lend money to buy gold, lend money against gold as collateral, and they can store the metal—all of which can generate worthwhile income for the banks as volume develops.

Group Ownership

Several vehicles, set up along investment company lines, have been proposed to enable investors to join together in buying gold on a group basis. Bars of Gold, Inc. is a bullion-owning company administered by the Calvin Bullock organization, which will sell and redeem its shares at net asset value, based on the daily London gold-fixing price.

Dreyfus Gold Deposits, Inc. and National Unit Gold Trust (the latter organized by National Securities & Research Corporation) are essentially gold purchasing and warehousing operations. American Gold Bullion Fund is still another investment vehicle, to be sold through the Bank of New York.

While the principle of group ownership may have some merit, because in some cases the investor can avoid assaying, fabrication and shipping fees as well as sales taxes, each of these plans should be carefully judged on its own merits. The tax status, minimum initial investment, and fees levied by the respective organizations are all matters to be carefully considered.

Principles to Follow

For prospective buyers of gold, in large or small quantities, this general advice is offered:

1. Check the reputation of the individual or organization through whom you are buying or selling gold—if there is any doubt, take your business elsewhere.

2. Make sure that any gold bullion you buy bears the mark of a known refiner, assaying weight, and fineness. *Ask for a written guarantee.* (En-

gelhard Minerals & Chemicals Corporation, for instance, is a well-known refiner which offers gold in bars up to 400 troy ounces, with its own stamp and in distinctive shapes.)

3. Keep in mind that the offering price to you will, of necessity, reflect a commission for the seller of gold. If the commission seems too high, it would pay to "shop around" for a better buy. A rough rule of thumb is that the commission or markup shouldn't exceed about 10 percent. Obviously you will pay a higher commission rate on small quantities of gold than on large purchases.

4. Be wary of any deals in which the seller stipulates that the customer is required to resell bullion back to him. This could mean that the price you may eventually receive will be determined by the seller, and his price could differ markedly from prevailing market quotations.

5. Be extremely cautious if you purchase gold bullion or coins on margin. You may be able to get in on as little as 10 percent margin, but you would be in an extremely vulnerable position in the event that the gold price declines and you are unable (or unwilling) to put up additional margin. Gold will be more—not less—volatile as the American speculator becomes a factor in the market place, and it's a safe bet that many margin buyers will be wiped out by price swings in the metal.

6. Remember that the U.S. Treasury will be keeping an eye on gold bullion transactions as a means of tax avoidance. The Government may have difficulty in keeping track of buyers and sellers, but don't count on it.

7. There are a number of unresolved questions concerning the tax status of investment vehicles which are designed to purchase gold in a group plan. If you go this route, seek qualified advice first.

8. Generally speaking, it is preferable not to take delivery of gold, except in small quantities. It is simpler and less expensive to let a refiner or broker store your gold and provide you with a document, similar to a warehouse receipt, attesting to ownership. Not only will you avoid red tape and expense, but also the probable assay cost on gold if it is in your possession when you want to sell it.

9. If you decide to accept delivery of gold, seek to do so outside of high-sales-tax states like New York (where an 8-percent sales tax applies on bullion transactions) because the sales tax can be a significant cost item. Your bank or broker can advise you in this regard.

10. If you are considering buying a very small amount of gold, say a 1/10-ounce wafer, our advice is to forget it unless you want it for a gift or memento. You will pay a hefty premium when you buy it, another sizable commission when you attempt to sell it, and this very likely will eat up any profit you could conceivably derive from an advance in the price of gold.

Chapter 4

Fundamentals of "Futures" Markets

"I'm very scared at what's happening now—the trend at present is leading to hell."

—Dr. Gunnar Myrdal,
quoted in October 1974
upon receipt of the
Nobel Prize in Economics

A futures contract is a legally binding instrument to buy or sell a designated quantity of a given commodity, at a specific time in the future, at a price agreed upon today. The contract will outline the standards that the commodity must meet to be acceptable for delivery.

Each commodity exchange has specific trading units, namely minimum standard quantities of the commodities it trades. For example, the minimum quantity in silver futures contracts on the New York Commodity Exchange is 5,000 ounces.

Typically, a commodity speculator is required by a commodity brokerage firm to provide a deposit or "margin" of approximately 10 percent to 20 percent before executing the order. But it is interesting to note that a speculator rarely lets the contract mature. It has been estimated that less than 3 percent of all contracts entered into in the futures markets are actually delivered.

What the speculator does is "cover" the original contract before it matures by assuming a position equal and opposite to the original trade. The liquidating trade has to be executed on the same exchange as the original contract and must be for the same delivery month.

Thus, if you *bought* 10,000 ounces of silver for delivery in December 1974, you would have covered by *selling* 10,000 ounces for delivery in December 1974.

Two Main Groups

Basically, there are two principal groups of participants who are active in futures trading: speculators and hedgers. The speculators (a classification that includes most of the public) are interested solely in price changes—they buy futures contracts if they think prices will go up and sell futures contracts if they think prices will decline.

The hedger, on the other hand, is usually involved in the production, processing, or marketing of a particular commodity and employs futures contracts as a tool for financial management purposes. In the case of gold, those who hedge would include miners, smelters, and fabricators of the metal.

Until the lifting of the forty-one-year ban on private ownership of gold in the United States, trading in gold futures has been strictly illegal for Americans.

But the ban ended on December 31, 1974, and interest in gold futures has been rising apace. A half-dozen American commodity exchanges are trading gold futures contracts, typically for 100 ounces of gold. (Spot contracts are for 400 troy ounces.)

On a standard 100-troy-ounce contract, worth about $18,000 (at a price of $180 an ounce of gold), you would be expected to put down a minimum "security deposit" or margin of about 8 percent, or roughly $1,440. But you most likely would have to put up a larger margin unless you are well known to a commodity brokerage firm—perhaps 16 percent. That is all to the good, because you would be in an extremely vulnerable position with an 8 percent margin. It's a good idea *not* to use all of your leverage when trading in a volatile commodity like gold.

Your costs would include your broker's commission (in the neighborhood of $45 on a "round turn" transaction), storage charges of between $2 and $5 per month for a 100-troy-ounce delivery, plus nominal insurance charges of about 50¢ per $1,000.

Now that trading in gold futures is a reality, such futures trading will probably become the most popular method of speculating in gold. And since the United States has more mature futures markets than Western Europe, not to mention the world's largest number of potential speculators, the United States will very likely become the world headquarters for forward trading of the yellow metal.

London will retain its preeminent position as the main "commodity" market for the metal, serving gold users with the well-oiled machinery, expertise, and tradition that enables the British capital effectively to handle large-scale receipts and deliveries of gold.

Zurich will continue to be a monetary center for gold, serving the needs and desires of those who are motivated by gold for its monetary aspects. It is suited for this due to the neutrality of Switzerland in world conflicts, its history of money-exchange handling, its nonpolitical protection of asset privacy, and its 800 years of monetary stability.

Considering that Americans have long been accustomed to dealing in a wide range of commodities (corn, wheat, soybeans, etc.) for future delivery, it seems unlikely that any major obstacles will arise to hinder the growth of forward trading of gold. In fact, some of the same exchanges on which futures trading in agricultural commodities is well established provide the market place for trading gold futures. So the United States should quickly catch up with London bullion dealers and Swiss banks that have been handling futures transactions in gold since 1954.

However, forward trading of gold is unlikely to follow exactly the same pattern as futures trading in more conventional commodities. For instance, there is little doubt that a fairly sizable percentage of forward buyers will take delivery of their gold when the contracts mature, assuming continued political uncertainty.

Those who are willing to assume a high degree of risk in return for large potential rewards will find the gold futures market a challenging and stimulating outlet for a portion of their speculative funds. However, some words of caution are in order. Remember that this is a highly volatile market, best suited for those who have ample risk capital at their disposal, as well as nerves of steel. Some knowledge of futures trading (or access to those with expertise in this area) is highly desirable. And be sure to know the people with whom you deal, as there is bound to be some risk of fraud. You don't want to wake up some day and find you are holding a bar of lead covered with a thin layer of gold!

The Winnipeg Market

Gold futures trading has been conducted by the Winnipeg Commodity Exchange since 1972. Some observers have contended that the failure of this market place (which for many years has provided a market for trading in contracts for future delivery of Canadian grains, oil seeds, cattle, and produce) to generate much excitement in gold may have negative implications for speculative interest in U.S. gold futures trading.

This conclusion, however, is not necessarily a valid one. The original Winnipeg gold contract covered the standard 400-ounce international bar; thus 400 ounces at $150 an ounce would mean an investment of around $60,000, and even at $100 an ounce it would involve $40,000 or so.

This contract apparently was too large to generate widespread investment or speculative interest, and evidently the Winnipeg Exchange came around to the same conclusion. It introduced a 100-ounce contract.

No use in trying to sell a Cadillac when the potential buyer is oriented towards a Chevrolet!

Exchanges Dealing in Futures

Among the exchanges actively dealing in gold futures are the following:

1. The Commodity Exchange (COMEX) in New York.
2. The Chicago Board of Trade.
3. The Chicago Mercantile Exchange (International Monetary Market).
4. The Midwest Commodity Exchange.
5. The Pacific Coast Commodity Exchange.
6. The New York Mercantile Exchange.

Details of contracts on four of these exchanges are provided in the accompanying table:

With the services provided by these and other exchanges, the would-be purchaser of gold futures should have no difficulty in obtaining access to the facilities he requires. With members of these exchanges, the major Swiss banks (as well as many smaller Swiss institutions), London bullion dealers, leading Canadian banks (such as the Bank of Nova Scotia), and others serve the individual's needs.

European banks, generally speaking, will be trading gold futures contracts while American banks will be restricted to dealing in actuals.

For those who understand the risk-reward ratios inherent in gold futures trading, and are willing to accept those risks, the following points should be considered:

1. The most attractive exchange for the investor is the one with the smallest size contract, the largest daily price change limit, and the greatest liquidity.

2. An investor should deal with a member broker of one of the exchanges dealing in gold, or with a broker who has a reciprocating agreement.

GOLD FUTURES CONTRACTS AT
FOUR MAJOR U.S. EXCHANGES

	Commodity Exchange Inc.	*New York Mercantile Exchange*	*Chicago Board of Trade*	*Chicago Mercantile Exchange*
Contract unit	100 troy oz. (5% tolerance)	32.151 tr. oz. (1 kg.)	96.45 tr. oz. (three 1 kg. bars)	100 troy oz. (5% tolerance)
Trading months	Feb., April, June, Aug., Oct., Dec.	Jan., March, May, July, Sept., Dec.	Jan., March, May, July, Sept., Nov.	Jan., March, June, Sept., Dec.
Deliverable grade	not less than .995 fineness—all exchanges			
Minimum non-member roundturn commission	$45	$30	$40	$45
Trading hours (Eastern time)	9:45-2:10	9:25-2:30	9:45-2:30	8:50-1:10
Sample minimum margin at Bache	$2,500	$800	$2,500	$2,500
Merrill Lynch*	$2,750	$1,000	$2,750	$2,750

Source: The New York Times.

*When price is $175 to $200 an ounce

3. Be sure, however, to pick a reliable broker with access to a variety of trading markets who is cognizant of all aspects of gold. It will cost you no more to do so, since the price of the metal for future delivery is expected to be based on the closing London gold price plus a premium that will be governed by the interest rates prevailing for the currency in which the transaction is made.

Incidentally, it is well to remember that regardless of where your futures contract is executed, an exchange itself does no business in the commodities that are traded by its members. Its main function is to provide a center or meeting place where buyers and sellers conduct their business and make contact with other parties involved in the movement and marketing of the commodities in which they are interested.

Effect on the Gold Market

What will be the effect on the gold market now that futures trading has begun?

The answer can only be an educated guess, but the long-term impact should certainly be a positive one for bulls of gold. Some of the yellow metal will have to be made available for those who take actual delivery of gold.

One member of the Chicago Board of Trade has estimated that within six months after trading has commenced there would be somewhere around 100 metric tons of gold in the warehouses of the nation's commodity exchanges. Although it may be on the high side, this figure is roughly equivalent to the amount of gold supplied by Canada and the United States *combined* in 1974.

The buyer of gold is likely to have more than a passing interest in foreign currencies. One should not forget that currencies with the highest gold backing (Swiss franc, Dutch guilder, and Lebanese pound) indirectly provide the purchaser with an interest in gold. In all probability there will be considerable shifting back and forth between gold and the world's stronger currencies by sophisticated traders and investors.

International currency futures contracts also will attract speculative interest. These are contracts that call for the purchase or sale of a specified quantity of a particular currency at a future date. When entered into by the gold trader, they can offer additional opportunities for profits, because of the arbitrage possibilities in conjunction with dealings for the metal.

For instance, assume that a one-year futures contract for gold is selling in London at the same level as a one-year futures contract in New York. At the same time sterling is selling at a sizable discount from the dollar.

It is possible to capitalize on this situation by setting up an arbitrage. In effect, you can buy gold with discounted sterling and thus acquire gold at a discounted price.

Use of Straddles

Another trading technique that may be employed is a "straddle," which is sometimes also referred to as a "spread." This involves the purchase of one future month against the sale of another future month of the same commodity. A "straddle" trade is based on a price relationship between the two months and a belief that the "spread," or difference in price, between the two contract months will change sufficiently to make the trade profitable.

Example: Assume that a person is extremely bullish on gold and sees near-term tightness in the market. In the case of a nonperishable commodity like gold, the difference in the cost of delivery between one month over another should, under normal market conditions, reflect the cost of carrying the metal—i.e., interest, storage, and insurance.

But, in a period of exaggerated near-term demand, the normal premium that reflects the carrying charges may disappear as buyers prove willing to absorb the carrying costs.

Under such circumstances spot gold could actually sell at a premium over forward months. Traders would likely react to such a situation by buying a nearby month and selling a forward month.

Where Currencies Are Traded

Following World War II the exchange rates among Free World nations were governed by the Bretton Woods Agreement of 1944, which helped to minimize currency fluctuations around fixed exchange rates. The world's central banks set up a system to keep exchange rates for the U.S. dollar within a one-percent range above or below a declared par value. But with the collapse of the agreement in 1971, there evolved a so-called "floating rate" system. This entailed much greater variability and volatility in rates, inasmuch as the central banks, under a floating rate system, do not protect either the upper or lower currency fluctuation limits.

As we noted in Chapter 2, this has meant serious risks for multinational countries and others engaged in international business operations. For example, if an importer bringing in goods from The Netherlands settles his debts through the purchase of Dutch guilders with U.S. dollars, and if the dollar should decline relative to the guilder during the transaction, he might wind up with a loss instead of a profit.

This is the reason that importers, exporters, financial institutions, and others must protect themselves against the risks associated with widely fluctuating "floating" exchange rates. International currency futures contracts can be quite useful in this regard.

There are several major markets for trading international currencies.

The International Monetary Market of Chicago was set up in 1972 to trade contracts in various kinds of currencies.

The Mercantile Exchange has set these minimum and maximum fluctuations in connection with the individual currency units:

	Minimum Fluctuation		*Maximum Variation*	
	Per Currency	Per Unit	Per Currency	Per Unit
British Pound Sterling	.05¢	$12.50	3¢	$750
Dutch Guilder	.004¢	$ 5.00	.6¢	$750
Deutschemark	.005¢	$12.50	.3¢	$750
Italian Lira	.00004¢	$10.00	.006¢	$750
Swiss Franc	.005¢	$12.50	.3¢	$750
Japanese Yen	.0001¢	$12.50	.006¢	$750
Canadian Dollar	.01¢	$10.00	.75¢	$750
Mexican Peso	.001¢	$10.00	.075¢	$750
Belgian Franc (commercial)	.0005¢	$10.00	.375¢	$750

Subsequently, in 1974, the New York Mercantile Exchange also instituted trading in futures contracts covering many of the world's leading currencies. The following contracts were in effect on the Mercantile Exchange in the latter part of the year and illustrate the trading possibilities.

Currency	*Unit*
Belgian Franc (commercial)	BFR 2,000,000
British Pound Sterling	£25,000
Canadian Dollar	CAN $100,000
Deutschemark	DM 250,000
Dutch Guilder	HFL 125,000
Italian Lira	LIT. 25,000,000
Japanese Yen	¥12,500,000
Mexican Peso	MN$1,000,000
Swiss Franc	SFR 250,000

It should be recognized that, together with the opportunities, there are sizable risks in trading foreign currencies. This activity, in fact, has been likened to "Russian roulette." It is well known that banks lost hundreds of millions of dollars during 1974 in playing this market, and they are supposed to be "experts" at it!

Still, for those individuals who are willing and able to assume the risks of fluctuating exchange rates in the hope of potential profits, the opportunities can be quite interesting. Consider, for example, that the Swiss franc traded at 0.3586¢ in May 1974, a 22.3 percent climb from its level in January of the same year (0.2931¢). This meant a potential 223 percent gain for a speculator.

Now assume that an individual believes that this trend would carry further and the American dollar would show more weakness against the Swiss franc. It would, therefore, be logical to sell dollars short and buy Swiss francs. This could be done by instructing a Swiss bank to purchase francs amounting to the equivalent of $100,000 for delivery in twelve months.

A customer would mail a check to the bank for 20 percent margin, or $20,000, that the bank would credit to a non-interest-bearing margin account. The bank would sell the client the Swiss francs equivalent to $100,000. The sale would take place at the twelve-month forward rate and the customer would pay slightly more than if he bought Swiss francs for immediate delivery.

On the due date the client can either cancel out the contract or he can roll over the contract for another period of one to twelve months. That is, assuming he has not sold back his francs to the bank before the due date of the contract.

This type of currency futures contract, in effect, offers the possibility of acquiring a claim for the equivalent of $100,000 in Swiss francs with a down payment of $20,000.

Incidentally, forward currency contracts can be obtained not only in Swiss francs, but in German marks, French francs, Spanish pesetas, Dutch guilders, or pounds sterling—providing, of course, that each contract exceeds $100,000 or its equivalent.

Another way of playing the shifts in foreign currency rates is through the purchase of three-month or six-month Swiss franc certificates of deposit at an Austrian bank where the going rates of interest are substantially higher than those prevailing in Swiss institutions for the Swiss franc.

These certificates of deposit (CDs) of Austrian banks can be denominated in any currency. Thus an individual who felt that the Swiss franc was due for an eventual rise could purchase a CD in that currency; in addition to the possible increase in value arising from the hoped-for increase in value of the franc, the purchaser would also receive a reasonably attractive interest return on his investment. Of course, if a currency loses value after purchase the holder will wind up showing a loss on the transaction.

Chapter 5

Role of International Banks in Gold Trading

"I have been accused of being worried over this inflation. I wasn't worried. I was just confused. . . .When you are worried, you know what you are worried about, but when you are confused you don't know enough about a thing to be worried."

—*Will Rogers*

More Americans than ever before are making nontraditional types of investments these days—gold coins, silver, gems, foreign currencies, stamps, works of art, and so on. The trend reflects not one but a number of influences, including investor expectations of the future course of inflation, disenchantment with the results of equity investments, and even a desire to avoid government impediments to the free transfer of their funds.

As they have become more sophisticated in their financial dealings, many Americans have opened Swiss bank accounts for the first time. The number is difficult to estimate. But it can be predicted with some certainty that, with legalization of gold ownership, many Americans will be encouraged to take advantage of the well-known Swiss flair for expert banking and financial services.

Their mounting interest is justified by past history. The performance of the Swiss franc against gold—or against virtually any other financial yardstick for that matter—has been exemplary for more than a century. While runaway inflation ravaged most European economies and hyperinflation destroyed the German mark in the 1919-1924 period, the Swiss franc remained immune to the monetary debauchery raging all around it and closely maintained its value against gold.

During the 1921-1934 period, when banks collapsed like tenpins all over Europe and more than 14,000 U.S. banks failed, only one minor Swiss investment bank became insolvent.

Opening An Account

Those who wish to open a Swiss bank account will find it a simple matter, depending, of course, on the particular bank. It should be emphasized that it is perfectly legal for any U.S. citizen to establish a bank account in Switzerland; all that is required is to indicate the existence of such an account on your U.S. income tax return.

The "Big Three' in the Swiss banking industry are the country's three largest institutions—the Union Bank of Switzerland, the Swiss Bank Corporation, and the Swiss Credit Bank. In addition, among the more than 400 Swiss institutions, there are a number of medium-sized and smaller banks with expertise in serving international clients. In the latter category is the Foreign Commerce Bank of Zurich and Geneva, which is reported to serve Americans efficiently who are frequently confused by red tape and language difficulties.

It is interesting to note that Swiss banks, generally speaking, do not solicit American customers—that is not their way of doing business. Nevertheless they have an impressive range of banking capabilities, including some that a good many American banks tend to lack.

For instance, some Americans have begun to invest in foreign currencies in much the same way that many professionals do—namely instructing their banks to roll over (or keep reinvesting) their funds in short term Eurocurrency time deposits or certificates of deposit, denominated in hard currencies like the Swiss franc, the mark, and the guilder.

Trading foreign currencies in this manner carries risks, but in times of currency turmoil it has proven an effective means of safeguarding—and even adding to—one's capital. Importantly, it provides diversification of assets.

Besides offering protection in a crisis, a Swiss bank can transfer your currency from country to country—just about anywhere. In fact many Swiss banking institutions issue check books that enable the drawer of the check to pay in any currency he may desire. This is extremely useful at times of fluctuating exchange rates. For instance, it can enable a traveler to pay his bills in the currency that offers him the most favorable rate at a given time.

Of course the Swiss banks also offer complete privacy in banking transactions, which an individual may find desirable from the standpoint of his personal and family relationships.

Unless you are a heroin dealer or an international swindler, you need have no concern whatsoever that anyone will tamper with the privacy of a Swiss bank account.

Swiss banking is business-oriented, and in no way politically motivated. This is a key to their success. Unlike North America and other European countries, Swiss bankers, from the central bank down, are career bankers, never political appointees, so they don't come under political pressure. There is no general manager of a big Swiss bank who is also a member of parliament.

Contrary to the belief held in some quarters, Swiss bank accounts normally pay interest—usually between 4 percent and 6 percent.[1] Although this rate may be lower than that obtainable in some U.S. money-market instruments, it has this additional feature: if the Swiss franc—widely recognized as one of the sturdiest currencies in the world—should appreciate against the U.S. dollar, the account holder would have the possibility of significant capital enhancement. Needless to say, there is also downside risk in such transactions.

Thus the Swiss banks can be useful to any actual or potential investor. They can provide expert counsel on foreign investments—including gold —both as a broker and an investment advisor. Their perspective incidentally is not limited to one market, but takes in the entire international investment scene.

It is true that certain other countries (such as the Bahamas, the Cayman Islands, Bermuda, Singapore, etc.) also offer numbered accounts. But the political and economic stability of these countries is not in the same league with that of Switzerland.

In Switzerland, banks are subject to stringent regulations. Cash liquidity requirements approximate 15 percent of total deposits and this total must be readily available for depositors. Banks, moreoever, can increase their loans by only 7 percent per annum. Such high cash liquidity, plus strict limits

[1]A penalty rate of interest was levied by the Swiss late in 1974, but is a temporary measure of currency control and applies only to new deposits.

on loan expansion, goes a long way toward insuring the stability and solvency of the Swiss banking institutions.[2]

Investors who are concerned about bank liquidity these days (and rightfully so) can check three main points in the financial statement of a domestic, Swiss, or other foreign banking institution:

1. As an indicator of cash liquidity, make sure that the bank's actual cash position and due-from-banks at-sight are higher than the due-to-banks at-sight item on the liability side of the balance sheet.
2. A healthy ratio should exist between collateralized and uncollateralized loans. Ask for details as to how the collateralized loans are in fact collateralized, to be sure there is no "window dressing" in the balance sheet.
3. Some banks have borrowed short-term funds and loaned out long-term money to their clients. This is a potentially troublesome situation and it would be revealed if the amounts due from clients appeared as a long-term item on the asset side of the balance sheet, and the amounts on the liability side showed only funds due to banks and to clients either at-sight or within a relatively short period.

Swiss Banks and Gold

We have already pointed out that although the United States most likely will emerge as the principal center for the trading of gold futures, the chief market for day-to-day trading in "spot" gold will remain in Europe.

It is a simple economic fact that the world's number-one gold supplier, South Africa, cannot ship large quantities of the metal to the United States as cheaply or as efficiently as it now ships to Europe. The shipping cost from Johannesburg to London is estimated at only 5¢ an ounce via the Union Castle Line, and that is cheaper than any competitor could possibly offer.

The sheer volume of South Africa's shipments requires that it be sold in a major world market. And the leading London gold dealers (such as Mocatta & Goldsmid, Sharps Pixley, and Samuel Montagu) have been in this business for well over a century. Experience helps, because the bullion business requires skill not only in shipping but in weighing, assaying, and insuring.

Consider for a moment the scandalous situation that exists at Kennedy Airport, where theft and pilferage are a way of life! That should convince anyone that the world's gold-trading center will remain in Europe for the forseeable future. Zurich is certain to remain the chief monetary center for gold. As pointed out, Zurich offers the complete anonymity and free-

[2]The banks mentioned in this chapter are located as follows:
Foreign Commerce Bank, Postfach 1006, 8022 Zurich, Switzerland; and P.O. Box 302, 1211 Geneva, 3 Rive, Switzerland.
Swiss Bank Corporation, Aeschenvorstadt 1, 4002 Basel, Switzerland.
Swiss Credit Bank, Paradeplatz, 8022 Zurich, Switzerland.
Union Bank of Switzerland, Bahnhofstrasse 45, 8021 Zurich, Switzerland.

Gold takes many shapes. Courtesy, Engelhard Industries

Gold and silver bars. Courtesy, Engelhard Industries

WORLD GOLD PRODUCTION
(In metric tons)

	1968	*1969*	*1970*	*1971*	*1972*	1973
South Africa	969.4	972.8	1,000.0	976.6	908.7	852.3
Canada	83.6	79.2	74.9	68.7	64.7	60.0
U.S.A.	46.0	53.9	54.2	46.4	45.1	36.2
Other Africa						
Ghana	22.6	22.0	21.9	21.7	22.5	25.0
Rhodesia	15.5	14.9	15.0	15.0	15.6	15.6
Zaire	5.3	5.6	5.5	5.4	2.5	2.5
Other	3.7	3.3	2.0	2.5	1.7	1.7
Total Other Africa:	47.1	45.8	44.4	44.6	42.3	44.8
Latin America						
Colombia	7.5	6.8	6.8	5.9	6.3	6.7
Mexico	5.5	5.6	6.2	4.7	4.7	5.0
Nicaragua	6.0	3.7	3.6	3.3	2.8	2.8
Peru	3.3	4.1	3.2	3.0	2.6	2.6
Other	11.2	15.3	15.6	17.2	18.5	18.9
Total Latin America:	33.5	35.5	35.4	34.1	34.9	36.0
Asia						
Philippines	16.4	17.8	18.7	19.7	18.9	18.1
Japan	7.0	7.8	8.4	7.7	9.6	10.4
India	3.6	3.4	3.2	3.7	3.3	3.3
Other	3.0	2.8	2.8	2.1	1.7	2.7
Total Asia:	30.0	31.8	33.1	33.2	33.5	34.5
Europe	6.7	6.9	7.4	7.6	13.2	14.3
Oceania						
Papua/New Guinea	0.8	0.8	0.7	0.7	12.7	20.3
Australia	24.3	22.2	19.5	20.9	23.5	18.4
Other	3.6	3.2	3.6	3.1	3.2	3.2
Total Oceania:	28.7	26.2	23.8	24.7	39.4	41.9
Free World Total:	1,245.0	1,252.1	1,273.2	1,235.9	1,181.8	1,120.0
U.S.S.R.:	304.2	318.2	335.5	344.8	360.2	370.6
Other Communist Countries:	8.4	8.4	8.4	8.4	8.4	8.4
Total World Production:	1,557.6	1,578.7	1,617.1	1,589.1	1,550.4	1,499.0

Source: Consolidated Gold Fields

THE BANK OF NOVA SCOTIA
TORONTO, ONT., CANADA

SAMUEL MONTAGU & CO. LTD.
LONDON, ENGLAND

DELIVERY BY
THE BANK OF NOVA SCOTIA
TORONTO, ONT., CANADA

DEUTSCHE BANK A. G.
FRANKFURT A/M,
FEDERAL REPUBLIC OF GERMANY

UNION ACCEPTANCES LIMITED
JOHANNESBURG,
UNION OF SOUTH AFRICA

Gold Certificate

Quantity__________Ounces

Serial No. TOR0000

THE BANK OF NOVA SCOTIA undertakes to deliver, subject to the conditions set out on the back of this certificate,____________________fine ounces troy, (being__________fine kilograms) of gold to____________________

____

upon the surrender of this certificate at its office in Toronto, Canada.

Whenever conditions permit, this certificate may be exchanged at the said office for a certificate issued by any one of the other banks named above in which such bank undertakes to deliver the said gold at its principal office in the named city. Such exchange will be effected at the rates for the respective certificates prevailing at the time.

THE BANK OF NOVA SCOTIA

SPECIMEN

SPECIMEN

DATE____________________

CONDITIONS

1. The liability to deliver gold to the holder of this Certificate will be discharged by delivering to the holder gold bars having a minimum of 995 parts of fine gold per 1,000 parts and totalling the number of fine ounces of gold for which this Certificate is issued, plus or minus 2% or 10 fine ounces, whichever is less, and debiting or crediting to the holder, at the rate of the most recent London fixing price per fine ounce prior to delivery, any difference between the number of fine ounces of gold so delivered and the number of fine ounces of gold for which this Certificate is issued. If this Certificate is issued for 400 fine ounces or a multiple thereof, delivery will be made in bars of between 350 and 430 fine ounces of gold each.

2. If the original holder of this Certificate wishes the gold to to be delivered to some other person, the form of the delivery instruction set out below must be signed by the original holder or, if there are two or more original holders and it is not provided on the face hereof that the gold is to be delivered to them jointly, by any one of the original holders. Subject to such verification as may be required, the gold will be delivered in accordance with the delivery instruction and no other instruction that may be received from the original holder or, if there are two or more original holders, from any of them, will be observed. The person named in the delivery instruction may not himself give any further instruction but may request the issue of a new certificate in his name by completing the exchange instruction set out below and, subject to such verification as may be required, a new certificate will thereupon be issued. All signatures on the delivery and exchange instructions must be guaranteed by a Canadian chartered bank or other guarantor acceptable to The Bank of Nova Scotia.

3. The Bank of Nova Scotia reserves the right to require four business days' notice for delivery of the gold.

4. Upon the delivery of the gold, the form of receipt set out below must be executed. If the gold is to be delivered to two or more persons and it is not provided on the face hereof or in the delivery instruction that the gold is to be delivered to them jointly, any one of such persons may surrender this Certificate by executing the form of receipt.

5. Samuel Montagu & Co. Ltd., Deutsche Bank A. G. and Union Acceptances Limited are under no liability to deliver gold under this Certificate.

DELIVERY INSTRUCTION

TO THE BANK OF NOVA SCOTIA:

Please deliver the within described gold to

Date____________________

Signature Guaranteed

EXCHANGE INSTRUCTION

Date____________________

TO THE BANK OF NOVA SCOTIA:

I represent that I am the person designated in the above delivery instruction and request that you exchange the within gold certificate for a new certificate issued in my name.

Signature Guaranteed

RECEIPT

RECEIPT IS ACKNOWLEDGED of the delivery, in accordance with the terms of the within gold certificate, of the gold to which the holder of the said gold certificate is entitled thereunder.

Date____________________

dom from regulation that is characteristic of Switzerland and that no other gold-trading center offers.

In addition to being simpler, it should be just as cheap to buy gold in Zurich as anywhere else because of the long familiarity of Swiss banks and bullion dealers in trading the metal.

Remember, too, that the Zurich market is a free auction market that is open all day. The London market, though excellent in many respects, is somewhat unique in that "fixing" of prices takes place only twice a day. Many people expect to buy gold at the last fixed price, but actually it can change in minutes after the fixing.

Furthermore, foreign exchange trading, arbitrage transactions, and other financial dealings that have a relationship to gold all can be carried out in Zurich with expertise for which the Swiss are famous.

And not to be overlooked is another fact: in Switzerland, there is no capital gains levy on profits from gold holdings that may be realized by nonresidents.

The fact is that a big trader in gold—a foreign potentate or Arab oil prince—will prefer to keep his gold with a Swiss bank rather than with a highly regarded American banking institution such as the First National City Bank. Other investors, surveying the possibilities that are available, may also come to the conclusion that Switzerland is not only a natural haven for "smart money" but also a logical trading market for gold transactions.

Generally speaking, the small investor may feel more comfortable with an American brokerage firm, while the more sophisticated and affluent type of investor, desiring both safety and secrecy, may prefer to tap the rich lode of expertise in gold and related areas of money management that is traditional among Swiss bankers.

How a Swiss Bank Handles Gold Orders

For those interested in the details of how Swiss institutions handle gold orders, this is the chronology of such transactions:

Gold orders are received by a Swiss bank by letter, cable, or telephone. As soon as the order arrives, the bank contacts the Zurich gold pool or a London bullion dealer and proceeds to purchase the gold for the client. Orders are executed immediately or, if received during the night, are executed at the opening in the morning.

Gold bars of varying sizes are available, but for investment purposes the standard 12.5-kilogram bar is the most suitable one. (While it is possible to purchase bars of small size—subject to a minimum order of ten ounces—they are proportionately more expensive due to higher manufacturing costs.)

The gold is stored by the bank for its clients and is segregated from the bank's assets. Therefore, should a bank fail, the gold holdings cannot be touched.

As soon as the purchase has been made, a confirmation is sent to the client; his account is charged with the total amount of the purchase and his custodial account is credited with the respective quantity of gold.

If a client wishes to sell the gold, he follows a similar procedure: namely, instructing the bank by letter, phone, or cable. The bank again approaches the Zurich gold pool and sells the indicated quantity. Once a sale has taken place, the account is credited with the amount of the proceeds of the sale, and the custodial account is charged with the respective number of ounces of gold that have been sold.

Gold futures contracts (see Chapter 4), both long and short, can also be arranged. The minimum contract size is fifty kilograms.

A cash margin deposit amounting to 50 percent of the total contract amount is required; the 50 percent must be deposited in cash and will be credited to a non-interest-bearing margin account. The margin must be maintained during the period of the contract, and the bank is entitled to call for additional margin should it become necessary.

The mechanics for gold futures are the same as for spot purchases and sales, the only difference being the price; on a purchase the client would pay a premium for the futures price and on a sale he would also receive the premium. Premiums vary according to market conditions, but in general they are based on the Eurodollar rate.

For each purchase of a futures contract, the bank issues a confirmation including the due date of the contract. On the due date the contract is automatically closed out and settlement takes place on that day. However, contracts can be closed before the due date, or they can be "rolled over" on the due date.

The basic difference between the outright purchase of gold bullion and a forward contract in gold is this: when the gold bullion is bought, the client is the owner of the metal; in a gold futures transaction, the client only has a claim against a gold account.

Canadian Banks

Historically, Canadian banks have played a well-defined role in satisfying the demand for gold from the Dominion's own citizens and from others who are attracted by the yellow metal.

Prior to the bill lifting the U.S. ban on ownership of gold, most of these banks were in the uncomfortable position, usually, of having to reject Americans wishing to buy gold, while serving the needs and desires of Canadian citizens and other non-American customers. They did not, of course, consciously aid U.S. citizens in flouting the U.S. ban against gold ownership. As one Canadian bank official describes the situation that existed when Americans were prohibited from owning gold: "If we knew they were Americans, we couldn't and wouldn't sell to them."

But Canadian banks, after all, are not a legal arm of the U.S. Treasury. Like their counterparts in Switzerland, they simply do not undertake the extremely formidable task of subjecting every gold purchaser to a searching inquiry.

The past procedure, as followed by one major Canadian bank has been quite simple. It has refused to sell gold to anyone with an American address, but officers of the bank have not insisted on seeing the passport of the purchaser.

Thus it is difficult—if not impossible—to estimate with any accuracy the extent of American gold holdings in Canada. The amount is probably substantial, but it is doubtful if it approaches the hoard of the metal believed to be squirreled away in Swiss banks on behalf of U.S. citizens.

Insofar as Canadians are concerned, gold has been legally bought, sold, exported, and imported for years without license or legal restriction in the Dominion. In this respect the Canadians part company with those who consider gold a "barbarous relic." Instead, many Dominion residents look upon the metal as a stable, internationally accepted form of security and they have accounted for a fairly consistent, though not overwhelming, demand for the metal at the windows of the leading Dominion banking institutions. Inquiries about gold have accelerated from time to time when "gold fever" grips the public.

Now that Americans are no longer barred from gold ownership, the Canadian banks should be well equipped to handle orders for gold and should capture some part of the business that is available. A survey indicates that charges by the Canadian institutions for their services are relatively moderate, and their bookkeeping procedures are uncomplicated.

As an illustration of what a gold buyer might expect, in 1974 when the metal was quoted at $178 per ounce, a customer at a Windsor, Ontario, bank paid $178.30 an ounce for a ten-ounce bar and $178.75 per ounce for a five-ounce bar. On top of this, he was charged a bank commission of one-half of one percent, plus a $5 charge for shipping the gold from Toronto to Windsor. Charges for shipping from Canada to the United States are higher, but are scarcely prohibitive.

The Bank of Nova Scotia is generally regarded as Canada's largest gold trader and it is believed that this institution purchases the major share of the Dominion's output of the metal. Because many of its customers prefer a document attesting to ownership of gold, rather than physical possession of the metal, the Bank of Nova Scotia—which sells at more than 900 branches throughout the Dominion—offers gold certificates denominated in multiples of 10 fine ounces or one kilogram and traded on the same basis as actual gold. The certificates may be considered as warehouse receipts against unallocated gold.

These well-known gold certificates, which have come into fairly wide use since the end of World War II, are redeemable by the issuing bank in Canada. The metal represented by the certificate may be sold at any time and the proceeds transferred into virtually any monetary unit. A transfer of ownership of a certificate also is a relatively simple procedure.

The certificates have been used as collateral for loans and it is possible to purchase them on margin if the buyer is deemed to be credit-worthy. The simplicity of these certificates makes them attractive to buyers who prefer to avoid the cumbersome process of taking physical possession of gold; while at the same time they have an instrument that can be used as collateral, sold, or transferred to another owner without difficulty. Of course similar instruments have been developed by American sources—such as the mechanism set up by various stock exchange firms that issue certificates of title to gold.

The Canadian Imperial Bank of Commerce—which probably ranks second to the Bank of Nova Scotia as a "gold" bank—has a one-kilogram (32.15 ounces) minimum for gold orders, and implements orders either by issuing gold certificates or by actual delivery. Although larger customers tend to prefer certificates, a spokesman for the bank states that customers are divided about fifty-fifty between those taking certificates and those requesting delivery. (The Canadian Bank of Commerce does not issue silver certificates.)

The Canadian Bank of Commerce expects to be competitive with other banks offering gold. However, it remains to be seen whether this bank, like other Canadian institutions, will go after the business aggressively or simply regard it as "just extra business."

The Royal Bank of Canada also has a one-kilogram minimum order. Its commission charge is reportedly about $15 on a kilo, plus a nominal shipping charge if the customer wishes the gold transferred from Toronto to his own bank. The purchaser, of course, may leave his gold in Toronto and pay a storage charge for the privilege.

Chapter 6

Gold Coins: Bullion vs. Numismatic Coins

"Never in the three decades of the International Monetary Fund and the World Bank has inflation posed a more universal threat to the world's economic and social system."

—Hans Apel,
Finance Minister,
West Germany

For thousands of years men have been collecting gold coins. Countless fortunes have been made speculating or investing in coins; and lives have been saved during war when coins have been exchanged for food, or even bartered for a human life.

Small wonder then that the same attraction to coins that prevailed in ancient times persists to this day. Millions of people in all corners of the globe collect coins for esthetic pleasure, as a store of value, as a hedge against the collapse of paper currencies, and as a semipermanent form of investment.

Gold coins, by their very nature, have some fundamental advantages over gold bullion and gold shares: they are easily portable, they can readily be concealed, and they can be purchased and sold without extensive record keeping. They are ideal investments for the small investor, and likewise have a useful role in meeting the needs of large investors. For those who wish to take physical possession of their gold holdings, it makes more sense to do so in the form of coins than bullion.

On the negative side, it must be recognized that coins—like bullion—are non-income-producing investments and therefore must appreciate in order to offset their carrying charges such as insurance, storage, shipping fees, and the like.

It is common knowledge that coins have frequently been used to transfer wealth from one country to another. In some countries they are a popular means of avoiding punitive taxation. We do not recommend them for such purposes, but it is a recognized fact of life that gold coins, in effect, are a private approach to the ownership of gold in many parts of the world.

While the timing may not be precise, gold coins tend to move directly with the price of gold bullion. This tendency, it may be noted, is also true of gold shares, but the latter tend to discount market expectations heavily.

U.S. Policy

Minting of gold coins in the United States was halted in 1933, and the Gold Reserve Act of 1934, which prohibited the private ownership of gold, also confiscated all gold coins held by Americans. An exception was made for coins with "recognizable numismatic value."

This was an ambiguous restriction and it was not rigidly enforced by the Treasury. In any event, many coins were retained by their holders and it has been estimated that a substantial volume of these gold coins found their way abroad. (Many have since become collectors' items.)

In 1954 the Treasury declared that all coins minted before 1933 would be presumed to be rare. This action, in effect, legitimized the ownership of gold coins and virtually eliminated the threat of confiscation.

With the legalization of gold ownership, there is no further impediment to purchase of all types of gold coins, in any amount, regardless of any arbitrary distinction by the Treasury. But it is still worth noting that a prohibition against gold ownership, once lifted, can be reinstated again at some future time. Congress, it may be recalled, has traditionally been ambivalent on the subject, often expressing opinions but usually leaving the final decision up to the President. If history repeats itself and a prohibition against gold ownership is reinstated, we have no doubt that large numbers of coins would be retained by their owners. It would be far more difficult for an owner of bullion to resist such a ban.

Two Distinct Categories

Fundamentally, gold coins may be divided into two basic categories:

1. Rare or numismatic coins that derive much of their value from their scarcity and that are sought by collectors. Prices of such coins are far in excess of their intrinsic gold value. Evaluations in the market place, which reflect their scarcity plus the condition of the item, are highly subjective.

2. Bullion or intrinsic coins, though not especially prized for their scarcity, sell at only nominal premiums over their gold bullion content. Since a purchaser pays a premium of no more than 10 to 15 percent over the bullion value, and they can be bought and sold in fairly large quantities without disturbing the market, bullion coins are a practical means for the average investor to take a position in gold.

One example of a numismatic coin is an uncirculated 1855 dollar of relatively scarce design minted in the early 1850s. The price history of this item is revealing. It had a catalogue value of $12.50 back in 1948. In 1974 the catalogue value had leaped to $1,500.

This is by no means a typical case, but it does illustrate the extraordinary appreciation potential in carefully selected numismatic coins that have a tendency to grow in value.

Such coins, of course, are not easily acquired. As Les Fox, Numismatic Director of Perera Fifth Avenue Inc., points out, "It could readily take a full month to purchase just ten choice brilliant uncirculated Barber Half Dollars, 1892-1916, at a reasonable price. Further, the acquisition of ten choice lower mintage dates in the Barber Half Dollar Series at reasonable prices would probably involve a three-to-six-month search."

By way of contrast, Fox notes that it would probably be possible to confirm an order for up to 1,000 pieces of common coins like the U.S. gold twenty-dollar Double Eagle at the daily market quotation within an hour.[1]

There is no doubt that over the past few decades, numismatic-type coins have enjoyed a significantly higher degree of appreciation than the more conventional bullion-type coins that are acquired largely for their metallic value, and it may be anticipated that the investor with ample capital and patience will, with proper guidance, uncover further opportunities for long-term enhancement in the numismatic group.

Coin expert Fox observes that, "Even when common gold and silver coins have appreciated in value, numismatic coins have done better at the

[1]Perera Fifth Avenue, a foreign exchange firm and a part of the Deak-Perera Group, offers a 24-hour "gold line" service. By calling 212-586-2175, in New York, one can obtain by phone the current price for popular gold coins and the latest London gold bullion price.

very same time." Fox adds that, "The sum of $50,000 that may be allocated to rare coins in a $1-million portfolio will have more impact on rare coin prices than the remaining $950,000 will have on common coin prices."

Be that as it may, there is little doubt that the numismatic coin field is a Cadillac-type market and there are far more owners of Chevrolets than Cadillacs. In fact, of the estimated ten million or so coin collectors in the United States, it is believed that perhaps one-fifth are numismatically oriented while the remaining four-fifths are primarily interested in common coins. (Obviously, there is a considerable degree of overlapping between the two investor groups.)

In general we feel that the art of selecting numismatic coins is a fascinating and potentially lucrative pursuit and we commend it to the individual who is affluent.[2]

But this is a specialty beyond the scope of this volume. We are primarily concerned with protecting the average person against inflation and currency debasement; and the vast majority of people in this group, we feel, should direct their attention to bullion-type coins.

There are about a dozen coins in this category and they sell at modest to hefty premiums over their bullion content. The following half-dozen are popularly traded bullion coins with which investors would do well to familiarize themselves.

Austrian or Hungarian 100-Corona

In late 1974 these were the lowest-priced gold coins in terms of premium above actual gold content. These coins contain 0.9802 troy ounces of gold and are sold from 4 percent to 8 percent over bullion value.

The 1908 Hungarian 100-Korona and the 1915 Austrian 100-Corona are being restruck by their respective mints and are available in virtually unlimited quantities. The Austrian coin has proved a bit more popular in appearance but there is no problem in disposing of lots of 100, 500, or even 1,000 coins within an hour at the prevailing bid price or slightly below. Swiss banks are generally ready to buy lots of 5,000 coins, although this transaction might take several hours to complete. One could probably realize a better rate by selling in smaller lots to different dealers, which is what a Swiss bank would do unless the coins were needed to fill an order.

Mexican 50-Peso

Considered by many to be more popular and attractive than the above restrikes, the 1947 Mexican 50-Peso is available in unlimited quantity. It

[2]For those who wish to delve deeper into the subject of coins we recommend:

How to Invest in Gold Coins, Donald J. Hoppe (Arlington House, New Rochelle, N.Y., $8.95)

High Profits from Rare Coin Investment, Q. David Bowers (Bowers and Ruddy Galleries, Inc., Los Angeles, Calif. 90028)

Golden Profits from Olden Coins, Ira U. Cobleigh (Goldfax Inc., 25 Park Place, N.Y., N.Y. 10007, $3.70)

contains 1.0256 troy ounces of gold and sells for 5 percent to 10 percent over bullion. Collectors and investors have shown increased interest in the dates that are not being reissued, notably those in the 1920s that sell for about $40 more per coin than the restrikes.

Mexican 50-Peso coins can be bought or sold in quantity as easily as the Austrian and Hungarian coins. The smaller investor should keep in mind, however, that not all dealers will make firm commitments to buy or sell 100-coin lots without a deposit or without first taking possession of the coins.

U.S. $20 Gold Double Eagles

Of the two types of U.S. $20 Gold Pieces (struck from 1850 to 1933), the "St. Gaudens" design is the more popular. They should be purchased in strictly uncirculated condition although, due to being handled in bags, even mint-state coins will show many scratches or "bag marks" from contact with other coins.

The U.S. Double Eagle contains 0.97 troy ounces of gold and because of its relative scarcity sells for a premium of 80 percent to 100 percent over bullion value. Double Eagles get strong when gold is strong, but also get weak when gold softens so it is more difficult to trade them in a fluctuating bullion market. Also there is not as exact a bid for these coins as for bullion restrikes, nor is there always an eagerness to purchase at the bid price in lots larger than 20 or 50.

"Liberty"-type Double Eagles are generally available at a slightly lower price than St. Gaudens type but are also harder to dispose of. It normally is worthwhile to pay an extra dollar or so to get handpicked coins.

British Sovereign

Containing just under a quarter-ounce of gold, the British Gold Sovereign continues to be a popular bullion coin although its 40 percent to 50 percent premium makes it a bit expensive.

Sovereigns have not been minted since 1931. The so-called "new" Sovereigns (Queen Elizabeth type) are just starting to find their place in the American market but they are considered collector's items and usually command an extra premium. One would have no trouble selling a lot of 100 or 500 Sovereigns, but there might be some resistance at the 1,000 level and a seller would probably have to settle for a discounted bid at the 5,000- or 10,000-piece level. Example: if a "bid" for 100 pieces were $5,250.00, the "bid" for 10,000 pieces might be $490,000 or about 5 percent less per coin.

South African Krugerrand

This is a newly minted coin that has exactly one troy ounce of gold.

The South African Mint has supposedly made plans to issue half-size and quarter-size Krugerrands after January 1, 1975, which should add to the growing popularity of this coin.

Krugerrands were first struck in 1967, which also makes these coins a collectable series. But the main interest in the Kruger is that it has exactly

U.S. $20.00 Double Eagle

Hungarian 100-Korona

Mexican 50-Peso

Bullion-type gold coins. Source: Perera Fifth Avenue, Numismatic Division

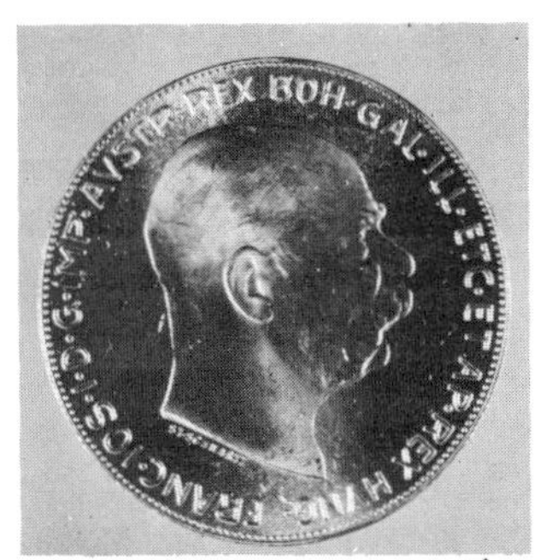

Austrian 100-Corona

South African Krugerrand

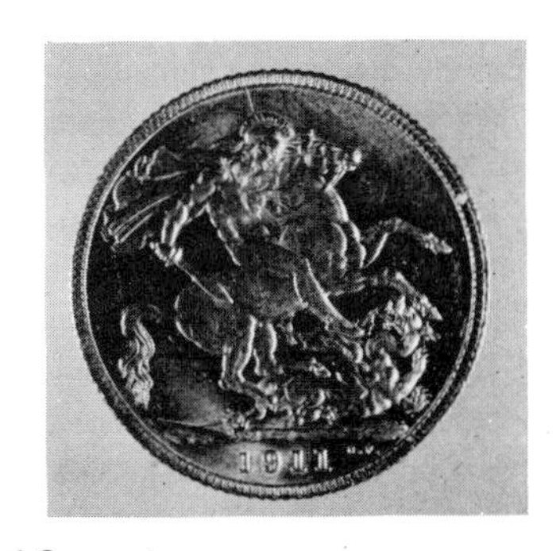

British One-Pound Sovereign

COMPARATIVE WEIGHTS AND METAL VALUE OF POPULARLY TRADED GOLD COINS

TYPE OF COIN	NET GOLD CONTENT (Troy Oz.)	ACTUAL VALUE OF GOLD CONTENT $120.00	$130.00	$140.00	$150.00	$160.00	$170.00	$180.00	$200.00	$225.00
U.S. $20 Gold	.9675	116.10	125.78	135.45	145.13	154.80	164.48	174.15	193.50	217.69
Mexican 50-Peso	1.2057	144.68	156.74	168.80	180.86	192.91	204.97	217.03	241.14	271.28
British Sovereign	.2354	28.25	30.60	32.96	35.31	37.66	40.02	42.37	47.08	52.97
Austria 100-Corona	.9802	117.62	127.43	137.23	147.03	156.83	166.63	176.44	196.04	220.55
Hungary 100-Korona	.9802	117.62	127.43	137.23	147.03	156.83	166.63	176.44	196.04	220.55

PREMIUMS ABOVE GOLD CONTENT AT VARIOUS SELLING PRICES @

		GOLD @ $120.00	GOLD @ $130.00	GOLD @ $140.00	GOLD @ $150.00	GOLD @ $160.00	GOLD @ $170.00	GOLD @ $180.00	GOLD @ $200.00	GOLD @ $225.00
U.S. $20	65%	$191.57	$207.54	$223.49	$239.46	$255.42	$271.39	$287.35	$319.28	$359.19
(both types)	90%	220.60	238.98	257.36	275.75	294.12	312.51	330.89	367.65	413.61
	120%	255.42	276.72	298.00	319.30	340.56	361.86	383.13	425.70	478.92
Mexico	10%	126.15	172.41	185.68	198.98	212.20	225.47	238.73	265.25	298.41
50 Peso	15%	166.38	180.25	194.12	208.00	221.85	235.72	249.58	277.31	311.97
	20%	137.62	188.09	202.56	217.03	231.89	245.97	260.44	289.36	325.54
Austria 100-C	8%	127.03	137.62	148.21	158.79	169.38	179.96	190.55	211.72	238.19
Hungary 100-K	14%	134.09	145.27	156.44	167.61	178.79	189.96	201.14	223.49	251.43
	18%	138.79	150.37	161.93	173.50	185.06	196.62	208.20	231.33	260.25
British	25%	25.31	38.25	41.20	44.14	47.08	50.02	52.96	58.85	66.21
Sovereign	40%	39.55	42.84	46.14	49.43	52.72	56.02	59.32	65.91	74.16
	60%	45.20	48.96	52.74	56.50	60.26	64.03	67.79	75.33	84.75

TROY WEIGHT EQUIVALENTS

1 Troy Ounce	= 31.103½ Grams
12 Troy Ounces	= 1 Troy Pound
32.15 Troy Ounces	= 1 Kilogram

one troy ounce of gold (the only bullion coin with that gold content) and many people (not to mention the South African government) are attempting to promote the coin as "real spending money."

Apparently there is a strong existing and potential market for this coin and many people are waiting for the opportunity to buy some. It should be interesting to see how many Krugerrands can be produced before demand levels off. If it becomes spendable money, demand may remain strong indefinitely.

Recommended Policy Toward Coins

As a generalization, it is our belief that approximately 10 percent to 15 percent of an individual's total portfolio be directed into coins. In special circumstances the percentage might be higher. For instance a wealthy individual who is not dependent on income and is primarily interested in the possibility of appreciation in coin investments might conceivably raise this ratio to around 25 percent.

As noted, we are not opposed to numismatic coin investment, but we feel that emphasis should be placed on bullion-type coins, notably the Mexican 50-Peso Centenario; the Hungarian 100-Korona; the Austrian 100-Corona; the U.S. $20 Gold Double Eagle; and the British one pound Sovereign.

We have been favorably disposed toward the South African Krugerrand, a bullion coin with "sex appeal." Of course there are other common gold coins that may be included in a coin portfolio but those mentioned are good representatives of this genre.

In view of the excellent marketability of bullion-type coins, there is no particular benefit to be derived from diversification per se. However some countries (such as Mexico) offer a wide variety of coins while others (like Canada) have struck relatively few gold coins. Thus some investors may wish to acquire several, rather than one, of the lowest premium common gold coins.

When it comes to numismatic-type coins, it is well to avoid being unduly concentrated in the coins of any one country, any one denomination, or a single mintage period.

Coin-Collecting Techniques

Collecting coins differs in a number of fundamental respects from purchasing gold mining shares, gold bullion or gold futures. In buying stocks, for example, the individual is usually doing business with a member of a major stock exchange or an over-the-counter house and the quotation he obtains on a security issue probably won't differ greatly from one broker to another. (Possible exception: in the over-the-counter market some brokers frequently provide better quotations, as well as execution of orders, than others.)

But unlike the stock brokerage business, the coin business is largely unregulated. There are, at the very least, several hundred important coin dealers in the United States and, including the one-man shops and fledgling firms, the total number may well run to several thousand.

Thus, in purchasing or selling coins, it is advisable to do business with a reputable concern such as a bank, foreign exchange dealer, member of a recognized securities exchange, or a coin dealer who has been in business for some time and is a well-situated member of his financial community.

If the dealer is a member of a recognized professional organization (such as the American Numismatics Association, the International Association of Professional Numismatists, or the Professional Numismatists Guild) so much the better.

Even if the individual has no doubts about the financial standing of the dealer, he should not hesitate to compare prices from several different sources —he might well wind up saving himself a fair amount of money.

It is well known that the number of coin dealers has proliferated in recent years together with the mounting interest in investing in coins. While most dealers are respectable businessmen, some have been known to engage in practices that are either illegal or unethical.

It is not unusual, for instance, to pass off "restrikes" as originals. (Restrikes are coins minted from dies of earlier years.) More serious is the risk of counterfeits being passed off as genuine. Counterfeits are a fact of life in the coin business; in Europe there are even quoted markets for both real and counterfeit sovereigns! However it is possible to minimize this hazard by following a simple practice: when you buy coins, ask for a guarantee of authenticity, *in writing*, from the dealer. If he is a legitimate operator, he will be only too happy to comply.

In early 1974 a Texas court, in a well-publicized case, enjoined one of the leading coin merchants from dealing in margin account contracts for the sale of gold and silver coins. It was alleged that a mere 3 percent of the margin account contracts held by the firm were actually backed or hedged by bags of coins; and while all margin account contracts of the firm were charged high rates of interest, it was alleged by the court that no loan identifiable with any investor could be segregated.

We recommend that the investor avoid dealing in any forward coin contract except in cases where he is thoroughly acquainted with what he is doing, and even then is conducting his business through a reputable bank or a member of an established commodity or securities exchange.

These are some additional points for the would-be coin investor:

- Acquire your coin portfolio carefully over a reasonable period of time rather than in a hasty, undisciplined manner.
- Be especially wary of "bait" advertising by dealers offering "fantastic bargains" or urging you to rush in to buy before it is too late.
- Give preference to uncirculated coins over worn coins, even though this may entail a modest premium. Quality is a significant factor in establishing the value of a coin.
- Beware not only of unscrupulous dealers but also of those who are under-

capitalized and who enter the field to make a "fast buck." They may wind up in financial difficulties, unable to serve their customers.

- Be wary of high-powered ads offering gold and silver medals as investments. The medallion mania has reached absurd proportions and medals are being struck to commemorate the most trivial events. If you collect medals as a hobby, fine. But don't buy them for investment purposes, because you may find them difficult or impossible to sell.
- Look out for treated or processed coins that are offered as uncirculated or "proof" items when buying numismatic coins. Also, diversify your holdings in this category—you may find that it is better to hold a reasonably wide variety of numismatic coins rather than one or two items of exceptionally high value.
- Remember that a coin collector faces a number of hazards, including counterfeiting, theft, and wear. Coins, therefore, should be kept in a safe place such as a private depository.
- Above all, seek out expert advice and look out for charlatans, especially if you are dabbling in the coin field for the first time.

Chapter 7

South African Gold Shares

"Gold is still our most important mineral and South Africa today produces about 78% of the total gold production of the Free World. Apart from gold, however, South Africa has been endowed with a wide variety of other minerals, including several of strategic importance."

—Hon. Gerald Browne,
Secretary for the
Dept. of Finance,
Republic of South Africa

No one seriously interested in gold can afford to ignore the South African gold shares; South African mines supply about three-fourths of the gold supply of the non-Communist world. And the South African gold mining companies have higher reserves, lower costs, and pay more generous dividends than competitors located anywhere else in the world.

Small wonder, then, that investors seeking a stake in gold without sacrificing income (which is inevitable with a sterile investment like bullion) turn to the "Kaffirs" or South African mining shares. Besides offering liberal income, they also can be rewarding holdings for their capital gains possibilities.

A Broad Spectrum

There are at least one hundred South African gold issues worthy of attention on the part of investors. They cover a broad investment spectrum, including speculative trading vehicles, so-called businessmen's investment situations, and sound income producers. All are traded in the U.S. over-the-counter market as well as in the London market.

The most common method for Americans to invest in this group is through American Depository Receipts or ADRs. (An ADR is simply a receipt for shares of stock in a foreign company.) For some years now, the average investor acquired stock in a South African mining company or "Kaffir" by buying an ADR in the over-the-counter market.

However, the interest equalization tax that was levied on foreign shares purchased by Americans is no longer in effect, and thus the professional investor may now find it more attractive to buy London-registered shares of stock. These often can be picked up at an appreciable discount from the price of the ADR when dealing outside of New York trading hours. Many brokers, in fact, will handle a transaction in precisely this manner and, by setting up an arbitrage between the ADR and the London-registered shares, can pick up an additional profit for themselves. If the investor avails himself of the services of a broker or trader skilled in taking advantage of such opportunities he may profit accordingly.

In addition to the investment diversity available among South African Kaffirs, the shares for the most part have good marketability plus highly predictable dividend income. In fact it may be said that South Africa's mining industry makes available more technical and financial data than just about any other industry in the world. At any given level for the price of gold, it is possible to work out the approximate life of a mine, the ores that will be mined, and their grades.

By making assumptions as to the rate of increase in costs, and at various levels of gold prices, the analyst can estimate the dividend that the mine can pay out over its entire lifetime. By the process known as discounting, we then can extrapolate the data so as to arrive at an approximation of the worth of each issue at a given price of gold.

In such a statistical undertaking it is well to point out that a number of factors—such as political risks, labor problems, floods, financial needs for expansion of mine facilities, even the sponsorship of a particular issue—also contribute to the market's evaluation and these cannot be computerized.

Still we do have a number of fundamentals to work with in appraising the investment worth of these issues. We know that gold mines are depleting operations. From the day a mine opens, it is pointing toward an eventual cessation of its operations. (The values being purchased by the investor consist of the ore in the ground, the knowledge of how to bring it out, and the equipment and the organization that accomplish this purpose.)

If a gold equity is to be considered attractive, we feel it must return to the investor, in the form of dividend income, the total cost of the share before the cessation of operations at the mine. Thus a mine with a ten-year life must yield (or be expected soon to yield) at least 10 percent when purchased. In actuality it should yield somewhat more. However, if the yield (or potential yield) is below the 10-percent level, very likely the investor will have paid an excessive price for the stock.

No discussion of South African gold shares is complete without considering the labor situation. Mining labor is presently derived from four locations—Malawi 28 percent; Mozambique 22 percent; Botswana, Lesotho, Swaziland 25 percent; and South Africa 25 percent. But the recent center of concern has been Mozambique. After four centuries of colonial rule, Portugal has granted conditional independence to Mozambique, with full independence to be official in June 1975.

Joachim Chissano, who is expected to be named premier of the new Mozambique government, has clearly stated that he would favor a policy of noninterference in the affairs of other countries. Concerning the South African policy of apartheid he said, "We will not be the saviours or the reformers of the South African policy. That belongs to the people of South Africa." Also to be noted is this fact: of the total foreign exchange that Mozambique earns, 66 percent is derived from South Africa and 16 percent from Rhodesia. (The new Cabora Bassa power plant project in Mozambique, incidentally, has one primary customer—South Africa.)

On this basis Mozambique appears to have two main options. It can continue its relations with South Africa (if not with Rhodesia as well), gradually diversifying away to new markets and new sources of revenue as structural changes are effected in the economy. Such a policy will result in a less revolutionary posture.

On the other hand, if Mozambique cuts off her links with South Africa and Rhodesia, it will almost surely suffer a critical economic crisis that will weaken the power structure of the new government. This is a most unattractive alternative. In all probability, Mozambique,[1] while professing coolness to South Africa, will be cooperative both politically and economically.

To sum up the political climate for South Africa, there is no doubt that the

[1]Mozambique receives for each worker provided outside of its borders a quarter of the worker's wage, paid to the government in gold at the artificial price of $42.22. This used to go back to Lisbon; it will soon represent a major source of financing for the new Mozambique government.

country's policy of racial separation will continue to be a lively topic in the halls of the United Nations, where the United States, Great Britain, and France have already vetoed a resolution to expel South Africa. The amalgam of Arab, African, and Communist countries in the UN will surely seek to make further political capital from the apartheid policy, striving for revolutionary, rather than evolutionary changes—even at the cost of South Africa's stability and solvency. This threat, we feel, will not succeed because South Africa, for the foreseeable future, is politically, economically, and militarily on a sound footing.

South Africa's Economy

For those living outside of South Africa, the Republic is considered virtually synonymous with gold. This is not surprising since it produces 78 percent of all Free World output of the yellow metal.

However, while gold is its most important mineral and (together with other minerals) contributes some 13 percent of the country's gross domestic product, it is worth noting that manufacturing industry accounts for 23 percent of the total domestic product.

South Africa produces some of the most advanced types of mining equipment in the world, manufactures diesel locomotives and other sophisticated products for both domestic and export markets, and is one of the most efficient producers of iron and steel in the Western world.

The nation's expanding technology is also reflected in its impressive strides in manufacturing petroleum products from coal—a development of no small importance to the United States—and in its role as a major supplier of uranium ore for atomic energy.

For the past quarter-century, the Republic's progress has permitted an annual growth rate of better than 5 percent per annum in real terms, which is comparable to that of Southern Europe and considerably higher than that of any other African nation except oil-rich Nigeria.

Leverage or Nonleverage?

South African gold shares may be classified into leverage and limited-leverage types.

Leveraged golds are those that have large amounts of low-grade ore and a high mill throughput (i.e., the ability to process ore). As the price of gold climbs, earnings and dividends of such mines can increase geometrically. However the risks are commensurately large.

In these situations the mines tend to be labor intensive—that is they require more workers to mine diminishing and lower grades of ore. For this reason they are vulnerable to increased costs and labor problems. Thus, for issues in this group to prosper, a rising market price for gold is critical. In our judgment, stocks in this category should be purchased only by those willing to accept higher-than-average risks in the hope of higher-than-average returns.

For the typical investor we prefer gold mining situations that are less labor intensive. Generally we favor a mine life of at least twelve years and a yield (actual or potential) of over 12 percent. Such mines, of course, would benefit handsomely from further gold price increases, but they would also suffer less in the event of a price decline. Furthermore, in most instances, these mines are substantial producers and, as such, would receive preferential treatment in the allocation of labor—should racial or political factors require a policy of this sort to be implemented.

Data On Individual Firms

In the following pages you will find pertinent data on twenty-four representative South African gold issues having a high degree of investor interest. Employing the present-value-discounted method, we have set forth an approximation of the theoretical worth of each share at various price levels for gold, together with the estimated dividend payouts.

It should be understood that this data, while useful, should not constitute the sole basis for investment decisions. Many other factors, including the price level of an issue at any given time, must be considered for intelligent decision making. Moreover, those with no previous experience in investing in this area should, if possible, avail themselves of competent advice from those familiar with the industry.[2]

South African golds constitute a volatile investment group and should be considered only by those in position to assume a significant risk factor. They are not for widows, orphans, or faint-hearted investors.

The individual companies follow, in alphabetical order:

BLYVOORUITZICHT GOLD MINING CO., LTD.

(Blyvoors)

issued capital —24,000,000 shares
dividend paid —February/August
located Far West Rand
group Barlow Rand
Gold and uranium producer

This medium-life mine (13 years) enjoys costs per ounce of ore produced of less than $50.00. It is not labor intensive but has some leveraged qualities. Blyvoors is presently mining ore at a grade of 16 grams per ton and it is expected that the grade decline into 1983 will be no more than a drop of 1.5 grams per ton to 14.5. On that basis Blyvoors will benefit handsomely from an increased price of gold. A new mill has been completed with a capacity of 160,000 tons per month. A uranium plant with capacity of 49,000 tons per month should add to company profits in the years ahead.

[2]To keep abreast of gold's ever-changing picture, as well as to obtain a perspective of the entire investment picture, the authors recommend a subscription to the *International Harry Schultz Letter,* at $200 a year, P.O. Box 1161, Basel, 4002, Switzerland; or Suite 226, 67 Yonge St., Toronto, Canada.

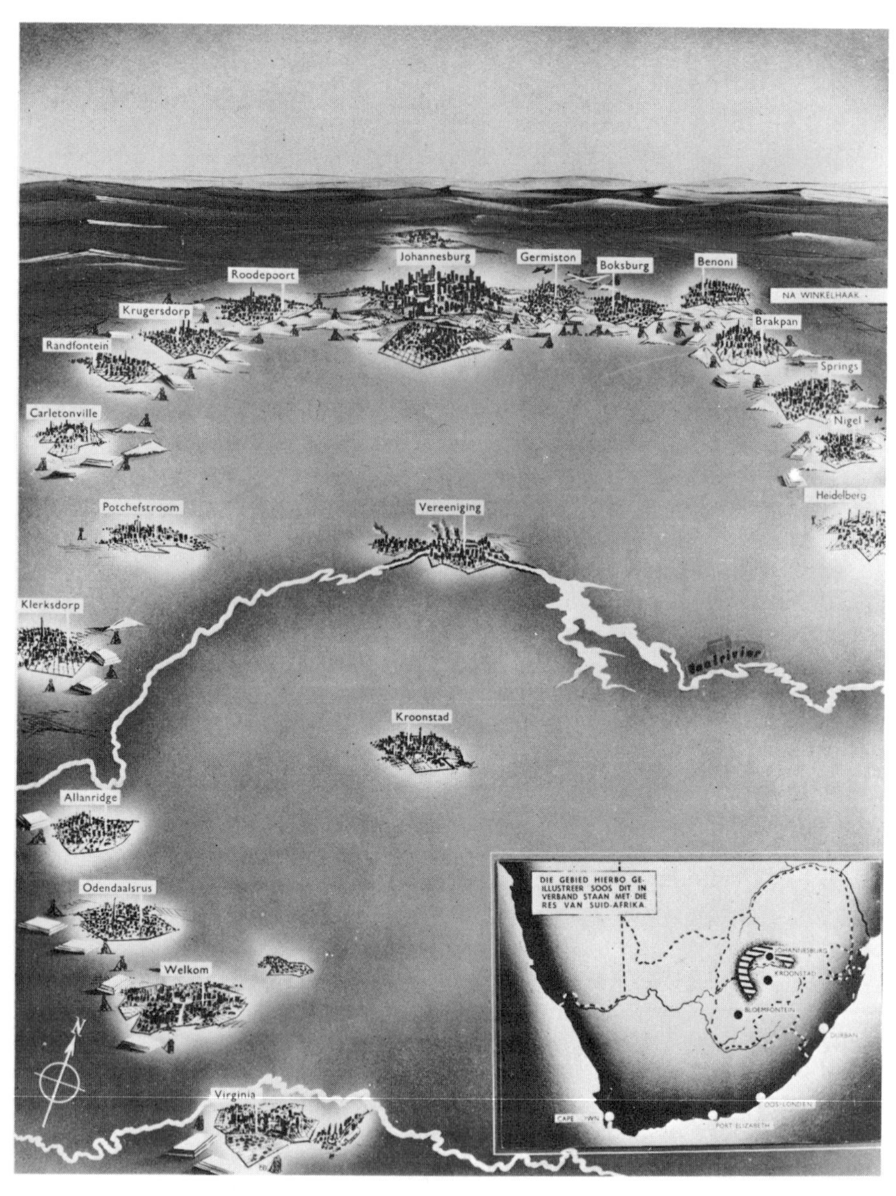

The Gold Arc in South Africa—location of the gold fields

Reduction works at Buffelsfontein gold mine, near Klerksdorp

Vaal Reefs No. 2 Shaft—Developing haulage towards Southvaal

South African smelter

gold price	$100	$150	$175	$200	$250
present value discounted 10%	9.00	14.50	18.00	20.00	25.00
estimated dividend	0.90	1.35	1.60	1.80	2.31

BRACKEN MINES LTD.

issued capital—14,000,000 shares
dividend paid—November/May
located Evander Area
group Union Corporation
Gold producer only

Bracken is not particularly labor intensive. It suffers from a short mine life regardless of the price of gold. A new area is under consideration by both Bracken and its neighbor, Leslie, that could result in life extension but this is only a speculative possibility. Faulting in this area makes the interpretation of the quality of sample bore holes difficult at best. Life of this mine is estimated between 9 and 11 years. Cost per ounce of gold produced is approximately $60.

gold price	$100	$150	$175	$200	$250
present value discounted 10%	2.90	4.95	5.80	6.59	8.35
estimated dividend	0.33	0.60	0.75	0.80	1.00

BUFFELSFONTEIN GOLD MINING CO., LTD.
(Buffels)

issued capital—11,000,000 shares
dividend paid—February/August
located Klerksdorp
group General Mining
Gold and uranium producer

Buffelsfontein has a high dependency on labor and has been plagued in the past by faulting problems. The life of this mine is estimated at 17 to 20 years. The cost of mining one ounce of gold is now $74.00. The relatively high grade of ore available from the new Orangia shaft should aid overall performance for the next two to four years. Uranium as a tail-end product will be helpful to Buffels. Increased mill throughput has aided results. Buffels is presently stockpiling ore ahead of its mill capacity to serve it during periods of labor shortages. These shortages can develop for nonpolitical reasons, when workers return to their homes to attend to seasonal agricultural pursuits.

gold price	$100	$150	$175	$200	$250
present value discounted 10%	14.00	27.00	36.00	40.00	56.00
estimated dividend	1.50	3.10	3.80	4.30	5.95

DOORNFONTEIN GOLD MINING CO., LTD.
(Doorne)

issued capital—9,828,000 shares
dividend paid—February/August
located Far West Rand
group Goldfields
Gold producer only

Doorne has a moderate dependency on native labor, but union problems and a shortage of European workers have caused production difficulties. The life of the mine is estimated at 19 years. Mining costs have been well controlled. The cost of mining one ounce of gold is presently $55. The grade of ore has dropped from 18.3 grams per ton to the present 12.7 grams per ton since 1970. It is interesting to note that Doornfontein is the mine at which new mining tools are being tested. In all probability Doorne will be the first mine to enjoy a new rock-cutting device that will reduce dependency on labor.

gold price	$100	$150	$175	$200	$250
present value discounted 10%	10.00	18.50	22.50	26.25	35.14
estimated dividend	0.78	1.60	2.02	2.25	2.98

DURBAN ROODEPORT DEEP, LTD.*

(Durban Deeps)

issued capital—2,325,000 shares
dividend paid—February/August
located West Central Rand
group Barlow Rand
Gold producer only

Durban Deeps is highly labor intensive. It is a leveraged mine by all criteria. It trades widely, offering both high risks and high rewards. Mine life is estimated at present gold price at 15 years. (Changes in the gold price in either direction will affect this estimate.) The cost of production of one ounce of gold is among the highest at $125. New promise in the Main and Kimberly Reefs may justify additional capital expenditures. Ore grades remain low, but at projected prices of gold overall earnings will be satisfactory. Durban Deeps represents a good leverage play with possible exploratory benefits.

gold price	$100	$150	$175	$200	$250
present value discounted 10%	12.00	33.90	39.00	42.30	44.60
estimated dividend	nil	2.80	3.10	3.70	4.12

EAST DRIESFONTEIN GOLD MINING LTD.

(East Dries)

issued capital—54,510,000 shares
dividend paid—February/August
located West Witswatersrand
group Gold Fields
Gold producer only

*Ultradeep mines in the "leverage" category tend not to conform to present-value-discount computations.

East Dries is not labor intensive. A favorite of both U.S. and European investors, it enjoys excellent marketability. The cost of mining an ounce of gold is among the lowest, at $27.40 by recent results. With the milling rate increasing towards a 1976 target of 181,000 tons per month, earnings are increasing. Life of this mine is conservatively estimated at 35 years. It is quite possible that ore grade may improve when the Carlon Leader stopes are opened. Because of the popularity of this mine and its long life the share price tends to exceed present valuation whenever gold shares are in general demand.

gold price	$100	$150	$175	$200	$250
present value discounted 10%	8.70	15.40	18.50	22.10	26.16
estimated dividend	0.22	0.49	0.69	0.81	1.26

EAST RAND PROPRIETARY MINES, LTD.*
(E.R.P.M.)

issued capital—3,960,000 shares
dividend paid—February/August
located East Central Rand
group Barlow Rand
Gold producer only

ERPM is a highly leveraged, high-cost producer that is critically dependent on the price level of gold. The cost of mining one ounce of gold is $108.20 at this labor-intensive mine. It operates more than two miles underground where the unusual geology of the depths can result in unexpected increases in costs at any time. The life of the mine at gold prices of $150 or more can be estimated at 15 years. Increased ore recoveries are being planned from more shallow levels in order to offset the ultradeep risks.

gold price	$100	$150	$175	$200	$250
present value discounted 10%	12.60	26.00	38.00	42.00	46.00
estimated dividend	nil	2.20	3.10	3.80	4.12

*Ultradeep mines in the "leverage" category tend not to conform to present-value-discount computations.

FREE STATES GEDULD MINES LTD.

(Free Gulls)

issued capital —10,000,000 shares
dividend paid —May/November
located Orange Free States
group Anglo American
Gold producer only

Free Gulls is a low-cost producer that is not labor intensive. The cost of mining one ounce of gold has been $31.10. This mine is a sound dividend producer and due to its joint ownership of Freddies with Western Holdings will enjoy fully the benefits of a higher gold price. Its shares trade conservatively, but participate well with any positive psychology in the gold market. Mill capacity is being raised to combat grade deterioration. Life of this mine is estimated at 23 years. During the latter part of mine life, ore grades could fall dramatically. This should be considered by the long-term investor.

gold price	$100	$150	$175	$200	$250
present value discounted 10%	23.00	36.00	39.00	55.00	66.20
estimated dividend	2.75	4.46	4.71	6.12	6.31

HARMONY GOLD MINING CO., LTD.

issued capital —26,884,650 shares
dividend paid —May/November
located Orange Free States
group Barlow Rand
Gold and uranium producer

Life of this mine has been extended to 20 years as a result of the merger of Virginia and Merriespruit with Harmony. Costs to produce one ounce of gold are high at $85.80. The mine is labor intensive. Combined milling rates are expected to be increased to 540,000 tons per month by 1976. Harmony should fare well with an increased price of gold. This situation combines some features of leverage with long life; as a result it has wide appeal to investors.

gold price	$100	$150	$175	$200	$250
present value discounted 10%	6.00	10.60	14.63	18.10	24.65
estimated dividend	0.41	0.95	1.30	1.48	2.01

HARTEBEESTFONTEIN GOLD MINING CO., LTD.

(Hartees)

issued capital—11,200,000 shares
dividend paid —February/August
located Klerksdorp
group Anglo-Transvaal
Gold and uranium producer

Hartees fully absorbed the leases of Zandpan Mines with a formula of 5.909 shares of Zandpan per share of Hartees. The tax carry-forward of Zandpan has been utilized by Hartees and the Zandpan Mines brought into full production. Expanded mill capacity, as a result of the merger, extends Hartees life by 2 years to 20. Increased prices of gold would have further benefits on estimated life. The cost of producing one ounce of gold is moderate at $62.30. The mine is reasonably labor intensive but is not fully leveraged.

gold price	$100	$150	$175	$200	$250
present value discounted 10%	18.25	36.31	43.92	49.41	53.43
estimated dividend	0.61	2.39	3.15	4.00	5.12

KINROSS MINES, LTD.

issued capital—18,000,000 shares
dividend paid —May/September
located Evander
group Union Corporation
Gold producer only

This issue has been neglected because of an accelerated rate of taxation and heavy capital expenditures involved in the opening of the new #2 shaft. The increased costs reduced revenue and required a drawdown on reserves to meet prior dividend rate. The cost of producing one ounce of gold is moderate at $55.45. The mine is not labor intensive. Substantial reserves of reasonable-grade ore exist. This, coupled with faulting that has resulted in ore relatively close to the surface for the Evander area, will serve to prolong life. With these reserves expansion of milling rate is possible. The life of the mine is estimated at 23 to 26 years.

gold price	$100	$150	$175	$200	$250
present value discounted 10%	4.23	7.21	10.21	12.02	15.31
estimated dividend	0.31	0.69	0.73	0.98	1.24

KLOOF GOLD MINING CO., LTD.

issued capital—30,240,000 shares
dividend paid—February/August
located Far West Rand
group Gold Fields
Gold producer only

With increased costs, onset of taxation, lower tonnage mined, generally higher costs, and a low price for its gold output Kloof's 1974 performance was generally unexciting. A combination of fires and pressure burst caused a shift in mining to lower grades. This will change as the situation is corrected and mining returns to the higher-grade areas. The negative results are reflected in the modest market evaluation of Kloof shares. The cost of mining one ounce of gold is excellent at $49.07. The mine is not labor intensive. After the opening of shaft #3 the milling rate will approach 3,000,000 tons. Life expectancy is long at 35 years.

gold price	$100	$150	$175	$200	$250
present value discounted 10%	15.40	19.60	23.61	28.70	32.51
estimated dividend	0.74	1.01	1.58	1.92	2.53

LESLIE GOLD MINES LTD.

issued capital—16,000,000 shares
dividend paid—May/November
located Evander
group Union Corporation
Gold producer only

At a $175-per-ounce gold price, mine life is expected to be 12 years. The cost of producing one ounce of gold is $76.20. Life expectancy of this mine can be influenced by developments in the northwest section of the lease; the company has stated it does not consider developments there could add more than five years to the enterprise. The company is not labor intensive and has demonstrated good cost controls in the face of general inflationary pressures. The shares pay substantial dividends and, assuming no labor problems, should return an investment well ahead of possible cessation of activities.

gold price	$100	$150	$175	$200	$250
present value discounted 10%	1.65	3.74	4.68	5.31	6.42
estimated dividend	0.20	0.51	0.61	1.00	1.42

LIBANON GOLD MINING., LTD.

(Libs)

issued capital —7,937,300 shares
dividend paid —February/August
located West Rand
group Gold Fields
Gold producer only

The life of this mine is 19 to 28 years, dependent on a price for gold from $150 to $250. The present cost of producing one ounce of gold is $47.33. The mine is not labor intensive. A new shaft has been vertically sunk and the milling rate is being increased to 140,000 tons per month. This, with announced additional prospecting area east of the northern half of the lease, explains the lively action of the shares in 1974. The relatively small float reduces marketability of the shares.

gold price	$100	$150	$175	$200	$250
present value discounted 10%	11.00	21.26	27.36	31.92	41.25
estimated dividend	0.78	1.56	1.98	2.31	3.31

LORAINE GOLD MINES, LTD.

issued capital —16,066,986 shares
dividend paid —May/November
located Orange Free States
group Anglo-Transvaal
Gold producer only

This is a situation highly geared to the price of gold; at prices under $150 only a small percentage of this property is viable. As the price of gold rises, more and more of its property becomes economic. The life of the mine ranges from 20 years at $100-gold to 40 years at $250. The cost of producing one ounce of gold is $92.90. The mine is labor intensive. In May 1974 Loraine announced plans to move forward on a major expansion; milling rate is being doubled to 200,000 tons per month. Full costs of expansion are programmed to be generated from internal sources. Dividends should remain at present rates as long as the gold price remains firm.

gold price	$100	$150	$175	$200	$250
present value discounted 10%	4.20	10.51	15.98	20.01	28.60
estimated dividend	0.34	0.71	1.00	1.15	1.41

PRESIDENT BRAND GOLD MINING CO., LTD.

(Brands)

issued capital—14,040,000 shares
dividend paid—May/November
located Orange Free State
group Anglo American
Gold and uranium producer

Brands was fortunate when it purchased Free State Saaiplaas, a played-out mine that was resurrected by the rise in the price of gold. It is also now prospecting to the northwest of the present mining lease. Even without any new developments, Saaiplaas has 11 years of life. The cost of producing one ounce of gold is $46.21. The mine is not labor intensive. Brands' own reserves have increased. The decline recently witnessed in ore grade will be arrested if gold remains firm. Number 4 shaft has disclosed good values at 15.8 grams per ton with reserves estimated at 1.5 million tons. Free State Saaiplaas will produce uranium for the life of President Brands.

gold price	$100	$150	$175	$200	$250
present value discounted 10%	19.00	38.20	48.27	52.63	71.50
estimated dividend	1.56	3.51	4.26	5.06	5.26

PRESIDENT STEYN GOLD MINING CO., LTD.

(Steyn's)

issued capital—14,000,000 shares
dividend paid—May/November
located Orange Free State
group Anglo American
Gold producer only

One man's good fortune can be another's hardship. The extension of life in Free State Saaiplaas (fully owned by Brands) has required Saaiplaas to fully utilize its own mill. Steyn had been intending to use that plant but now must rely on its own. Steyn appears to be a play on whatever new values can be received from #4 shaft. This shaft was negotiated from Sentrust-Lydenburg Plats General Mining for a 20-percent share of profits derived therefrom. The cost of producing one ounce of gold is $55.85. The mine is not labor intensive. Life of this mine is presently considered to be 18 to 20 years.

gold price	$100	$150	$175	$200	$250
present value discounted 10%	14.92	31.50	39.00	44.58	61.32
estimated dividend	0.82	1.60	2.10	2.28	3.41

THE RANDFONTEIN ESTATES GOLD MINING CO., LTD.

(Rand)

issued capital—5,413,553 shares
dividend paid—not yet initiated
located West Rand
group Johnies Consolidated
Gold and uranium producer

Randfontein witnessed a resurrection as the old played-out mine was resuscitated by the rise in price of gold. The new Cooke section is developing better than previous expectations. Labor intensity and cost per ounce of gold produced is not reflective of the true position, as Randfontein is a developing mine and not yet a fully projected capacity. Randfontein's life is estimated between 25 and 33 years dependent upon the gold price. The following are the earnings per share that can be produced at gold $150 at various rates of milling under the present ore grade:

90,000 tons per quarter	$2.91
210,000 tons per quarter	6.11
315,000 tons per quarter	9.17
420,000 tons per quarter	12.22

Uranium has shown interesting potential values in the gold section of this mine.

gold price	$100	$150	$175	$200	$250
present value discounted 10%	24.00	46.50	63.14	69.80	81.50
estimated dividend *	2.10	3.26	4.18	4.76	6.10

*Dividends are estimated for the various prices of gold at least 18 months into the future, at which time the mine should be functioning at full mill capacity.

ST. HELENA GOLD MINES, LTD.

(Saint's)

issued capital—9,625,00 shares
dividend paid —May/November
located Orange Free State
group Union Corporation
Gold producer only

Saint's produced one ounce of gold for $39.60. It is not a labor-intensive mine. The fact that the mine is shallow has resulted in good cost control and reliability. These, coupled with the steady grade of ore mined, have resulted in this issue gaining a preferred position among institutional gold investors. Encouraging results have been returned from Saint's exploration of Jurgenshof. The life of the mine is 25 years.

gold price	$100	$150	$175	$200	$250
present value discounted 10%	32.50	46.21	56.48	64.18	74.21
estimated dividend	2.63	3.21	4.01	4.37	5.38

SOUTHVAAL HOLDINGS, LTD.

(Southvaal)

issued capital—26,000,000 shares
dividend paid—nil
located Klerksdorp
group Anglo American
Gold and uranium producer

Southvaal is close to Vaal Reefs; it has a royalty agreement with Vaal Reefs that yields 55 percent of operating profits to Southvaal. The company is favored by a unique tax arrangement whereby its rate is fixed at no higher than 42½ percent. Such a tax level will redound to the benefit of profits as Vaal Reefs exploits their properties. In time, uranium will play a big role for Southvaal. Mine life is 35 years.

gold price	$100	$150	$175	$200	$250
present value discounted 10%	13.10	19.38	26.50	31.00	39.83
estimated dividend	0.26	0.30	0.42	0.61	0.70

VAAL REEFS EXPLORATION AND MINING CO., LTD.

(Vaals)

issued capital—19,000,000 shares
dividend paid—February/August
located Klerksdorp
group Anglo American
Gold and uranium producer

Vaals is the giant of South Africa, as well as the biggest gold mine in the world. Its shares are a barometer of the gold market. The cost of producing one ounce of gold is $63.90. The life of the mine is estimated at 35 years. Additional properties will come on stream as the gold price rises. Present heavy capital expenditures represent the exploitation of Southvaals property under royalty agreement with Vaal Reefs South. It is not labor intensive.

gold price	$100	$150	$175	$200	$250
present value discounted 10%	35.10	51.60	64.60	76.80	106.81
estimated dividends	2.10	3.51	4.12	5.01	6.23

WEST DRIESFONTEIN GOLD MINING CO., LTD.

(West Dries)

issued capital—14,082,160 shares
dividend paid—February/August
located Far West Rand
group Gold Fields
Gold producer only

The $27.40-per-ounce cost of production at West Dries is among the lowest in South Africa. The mine is not labor intensive. Life is estimated at 18 years. There is a considerable amount of low-grade ore in this mine that is economic at a gold price of $140 and above. The present ore grade of 27 grams per ton will not last the life of the mine but should produce excellent results for the foreseeable future. West Dries has uranium values but these will not contribute earnings over the near term.

gold price	$100	$150	$175	$200	$250
present value discounted 10%	44.40	63.28	71.97	79.36	101.98
estimated dividend	4.10	6.21	7.20	8.01	9.05

WESTERN DEEP LEVELS, LTD.

(Deeps)

issued capital—25,000,000 shares
dividend paid —February/August
located Far West Rand
group Anglo American
Gold and uranium producer

Deeps is a favorite of Johannesburg but its popularity has not spread to North America. The life of the mine is long at 35 years; the cost of producing one ounce of gold is $51.30. The mine has a solid dependency on labor but is not classified as labor intensive. Deeps should enjoy benefit from the proposed new mine of the Anglo American group, Buffelsdoorn; this mine will adjoin the southwestern corner of the lease. That will result in an increased minable area and benefit Deeps thereby. Profits from uranium production are presently unimportant.

gold price	$100	$150	$175	$200	$250
present value discounted 10%	24.56	36.74	45.24	51.82	61.98
estimated dividend	1.53	2.12	2.84	3.10	3.67

WESTERN HOLDINGS LTD.

(Holdings)

issued capital—7,496,376 shares
dividend paid —May/November
located Orange Free States
group Anglo American

Life of this high-grade mine is estimated at 18 years; its cost of producing one ounce of gold is $40.30. The mine is not labor intensive. Joint ownership of Freddies with Free States Geduld will add life to this mining operation. Ore grade will decline as time passes, but a firm gold price will compensate for this situation. This mine is considered investment grade by Johannesburg.

gold price	$100	$150	$175	$200	$250
present value discounted 10%	44.62	66.19	84.61	92.61	115.41
estimated dividend	4.11	6.34	8.51	9.58	11.63

Chapter 8

North American Gold Shares

"All major crises have been caused by previous inflation, which is far from being isolated and which, sooner or later, leads to collapse."

—Dr. F. A. von Hayek,
Co-winner of 1974
Nobel Prize in Economics

In the past, gold production in North America tended to flourish in hard times and wilt in years of general prosperity. Gold mining, in fact, was most profitable during periods of deflation. Consequently the shares of companies in this industry often moved in the opposite direction to the industrial stock price averages as well as commodity price levels.

One only has to look back to the onset of the Depression of the Thirties. As production costs fell, world output of gold turned sharply upward, rising from 19.2 million fine ounces in 1929 to 22.3 million ounces in 1931.

Under such conditions the purchase of gold issues was a satisfactory hedge against a receding economy. American gold producers—in contrast to most other depression-battered industries—experienced no deterioration of demand for their output or pressure on prices because the U.S. Government always stood ready to buy gold at a fixed price.

Unfortunately, in more recent times the fixed Treasury buying price of $35 an ounce for gold became an Achilles heel instead of a blessing for the industry. It left gold producers with no effective means of offsetting the inflationary influences of the post-World War II period. With steadily rising costs of labor, which represent somewhere between 60 percent and 70 percent of the industry's operating costs, the gold mining industry was severely squeezed between rising costs and fixed prices.

Although it was widely recognized that the industry was suffering from a prewar price level and a postwar cost structure for its output, little was done to help American gold producers. Canada, on the other hand, did establish a program of subsidy payments aimed at assisting producers at marginal mines and preventing mine employment from being slashed too sharply.

Perhaps one reason for the lack of sympathy for the gold miners is the fact that in the United States much of the gold produced was, and still is, a sideline by-product at mines whose prime function is to produce other metals such as copper and zinc. If it had not been for this by-product relationship, there is no doubt that a great deal of the yellow metal would never have left the ground.

Of course the dog days for the yellow-metal men are over. Thanks to the more realistic valuation of their product in world markets, gold producers are experiencing prosperous times that could continue well into the future.

Mines Are Reopening

The soaring price of bullion is stimulating exploration and reawakening interest in mines that have lain dormant for many years. Interest probably will be well sustained while bullion prices remain buoyant, but uncertainties surrounding most new and unproven ventures remove the securities of such companies from consideration as sound investments.

In this connection we would consider cautiously the often-extravagant claims of promoters of speculative gold mining ventures in the United States and Canada that have been given a new lease on life in the wake of higher bullion prices. Most of these projects are not on a sound footing from an engineering point of view and can be placed in the same category as a private treasure hunt.

Actually the gold mining industry in the United States is relatively small —it has been estimated that it ranks below the manufacture of cottage cheese in annual dollar volume.

The biggest domestic producer by far is the Homestake Mining Company mine at Lead, South Dakota. Also important are two Nevada properties—the Cortex mine operated by Placer Development Ltd. and the Carlin mine of Newmont Mining Corporation. Most of the rest of U.S. gold output comes as a by-product of silver and copper mining.

When an investor looks for a U.S. gold miner with a proven track record, he will find that Homestake Mining Company is the only American gold producer listed on the New York Stock Exchange. It is also one of the few North American golds that meet the investment yardsticks of demonstrated earning power, good marketability, sound finances, and proven dividend-paying ability.

The range of investment possibilities widens somewhat with the inclusion of several Canadian gold miners that, like Homestake, are listed on the New York Stock Exchange. Most prominent in this category are Dome Mines and Campbell Red Lake Mines. But here too it is evident that the number of situations suitable for general investment purposes is quite limited.

To be sure, the American Stock Exchange lists several gold producers but, generally speaking, we consider them as limited in their investment appeal though they seem to have a fairly wide market following. Giant Yellowknife Mines, for instance, is a fairly sizable Canadian producer, but it has an unimpressive record. Golden Cycle Corporation, another Amex company, has announced plans to resume gold production on its acreage near Cripple Creek, Colorado. But this firm is active in land development, movement of household goods, and transportation and, in our opinion, it should be judged as an industrial company rather than a gold miner.

Among the lesser-known "junior" Canadian golds, there are a number of possibilities including Camflo Mines Ltd., Agnico-Eagle Mines, and Sigma Mines. (Agnico is discussed later in this chapter.)

Camflo (Toronto Stock Exchange) is regarded as one of the more successful "junior" mines in Canada. Gross value of bullion production jumped to $10.5 million in 1973 from $5.9 million in 1972, and net income rose to $1.31 per share against 60¢ a year earlier. The average gold price received soared to $107.39 per ounce in 1973 from $59.77 in 1972, $41.19 in 1971, and $36.64 in 1970.

Camflo, incidentally, had an unusual item to report to its shareholders in an interim statement in 1974. According to the report, "early in the morning of April 1, 1974, three armed men forced their way into the mine at gunpoint and stole some 500 pounds of precipitate from the presses containing 2,462 ounces of gold worth approximately $416,700." The loss, the company hastened to add, was "covered by insurance."

Sigma Mines Ltd., another Toronto Stock Exchange listing, shapes up as one of the more interesting situations in the "junior" Canadian golds. The Sigma mine, located at Boulamaque, Quebec, has reported ore reserves of 1.2 million tons of 0.22 gold. This is sufficient for about two years of production at the recent rate of output, but chances seem good that new ore will be found. Sigma (controlled by Dome Mines) has increased earnings considerably since 1971. In 1973, Sigma reported net income of $1.82 per share as against 67¢ in 1972, and another sizable gain is indicated in 1974.

Limited Choices

It thus becomes apparent that the choices available to the investor in North American gold shares is extremely limited by comparison with the numerous opportunities in South African golds. What's more, it will be found that the South African gold mining shares sell at lower price-earnings ratios and offer more generous yields than North American gold shares.

In addition to these advantages, the investor in South African Kaffirs is able to obtain a virtually undiluted stake in the metal in which he is most interested—gold—whereas in most other areas this is usually not the case because companies are active in other fields besides gold.

This does not mean that opportunities in North American gold issues should be ignored. We feel that a balanced portfolio should include both North American and South African gold mining companies. However it is virtually impossible to ignore the South African golds in any well-rounded investment program oriented toward gold, and this is true whether the investor uses the direct approach of investing in gold shares or the intermedium of investment companies.

Investment Companies

There are also several U.S. investment companies heavily oriented toward gold, mainly South African golds. The reason for this interest in South African golds is because institutional money managers, like individual investors, have been severely limited in the number of domestic and Canadian gold producers that qualify for investment purposes.

One of the best known of the gold-oriented investment companies is ASA, Ltd. (formerly American-South African Investment Co.), which trades actively on the New York Stock Exchange. ASA has extensive investments in a portfolio of leading South African gold mining enterprises. Its management is in a good position to evaluate South African gold issues because it is closely interlocked with Anglo American South African, a foreign investment company that ranks among the world's largest mining trusts, and with Engelhard Minerals & Chemicals, a major U.S. dealer in precious metals.

Newmont Mining is another Big Board-traded investment company with a management that has been adept at mining investments, but unfortunately gold represents a minuscule portion of its variegated interests that generally encompass nonferrous metals and oil.

In the "open-end" or mutual fund category (Newmont and ASA are "closed-end" or fixed capitalization investment companies). International Investors Inc. has done well by tying its fortunes to gold. Since 1968 it has concentrated its investments in gold mining—mostly South African shares—though it also has investments in other industries. This fund offers shares to the public at net asset value plus a sales charge equal to 8¾% of the public offering price on sales of less than $25,000.

Another gold-oriented mutual fund that has done well in recent times is United Services Fund of Texas. This moderately sized no-load gold fund has holdings in South African gold shares.

An Overview

Taking an overview of the North American gold mining industry, we conclude that the outlook at current and prospective prices for gold is excellent. However the industry will remain exposed to some basic problems.

The major problem is the continuous rise in labor and operating costs, which, at least for the present, is offset by higher bullion prices. Extremely large capital expenditures, faced by both North American and South African mines, may also prove burdensome.

Another negative earnings factor is a general tightening up of antipollution standards that has required mines to dispose of their "tailings" in a manner that does not harm the environment. (Less stringent pollution standards in Alaska are a prime reason that Alaska is perhaps the most active gold mining state in the country.)

Finally, it is axiomatic in the mining business that, as the price of metal rises, it is sound policy to mine lower-grade ore. This factor may tend to prevent the impact of rising bullion prices from being fully reflected in company earnings.

In both Canada and the United States the main problem facing the industry is the scarcity of gold. The Dominion has remained the second-ranking producer of gold in the Free World but its reserves are being steadily depleted. Even if new ore is found, Canadian gold will remain difficult and costly to mine. (The same, incidentally, is true of Russia, which also suffers from difficult terrain and climatic extremes. But the Soviet Union apparently places a high priority on the production of gold, which is used to build up foreign exchange reserves needed to purchase products from the West, as illustrated by the huge 1973 grain deal with the United States.)

Despite the generally favorable outlook, a few words of caution for the gold share investor are in order.

The investment experience of the past several years indicates that prices of gold shares generally reflect advances in bullion prices. But this correlation is not necessarily precise for any given period of time.

It is also well to remember that, regardless of bullion prices, the all-important factors in gold mining are the estimated life of the reserves and the quality of the ore, together with the projected cost of getting the gold out of the ground. It will be of little comfort to an investor if the price of bullion soars into the stratosphere and he finds that he has invested in a mine nearing exhaustion.

It should also be recognized that prices of gold stocks fluctuate widely, often for no apparent reason. There was a flurry of selling in September 1974 when a well-known investment advisor suddenly recommended to his clients to liquidate their gold shares. This produced price declines that clipped as much as 50 percent from the market values of individual gold issues. However, a rapid recovery phase soon followed that eliminated most of the earlier paper losses. It is interesting to note that during this hectic market phase (which was characterized as a "liquidity panic" because of the high incidence of margin selling) gold bullion lost only about 3 percent of its value.

Timing, of course, is always a factor to be considered in the investment field, and the question might reasonably be asked: Is it too late to consider American or Canadian gold shares?

Based on historical yardsticks gold stocks do not appear to be selling out of line with the evaluation normally accorded this group. In fact, price-earnings ratios of about fifteen times (based on estimated 1974 earnings) for Homestake and Dome are rather low historically.

Even though we are favorably disposed toward the limited number of choices available in North American golds, we urge that caution, careful selection, and diversification be observed by the investor. Diversification is a "must" because any single mine may have floods, cave-ins, fires, or other catastrophes—which means you should avoid putting all of your eggs in one basket.

In addition we recommend that commitments in this group should be made strictly on a long-term basis, in order to avoid being whipsawed by short-term market swings.

We are providing further highlights of four companies that offer a cross-section view of the North American golds. This quartet, which covers a wide quality range, includes Homestake, Dome, and Campbell Red Lake in the investment category; and Agnico-Eagle, a speculative situation with growth potential. All of these issues are quite volatile and we recommend that their prospects be analyzed in the light of prevailing market conditions before any investment decisions are made.

HOMESTAKE MINING COMPANY

Homestake, the largest American gold producer, has operated profitably for nearly a century. For many decades it has operated a mine at Lead, South Dakota, that ranks as one of the world's largest gold properties.

In recent times Homestake has been confronted with the problem of maintaining its earning power in the face of mounting costs for labor, equipment, and supplies. These conditions resulted in a lackluster earnings performance during the decade ending in 1972. But even so the firm remained profitable with the help of continued improvements in production techniques, as well as the use of new equipment. It has also maintained an exceptionally strong financial position, with cash items alone far in excess of total current liabilities.

In 1973 ore production from the Lead, South Dakota, mine was 1,573,763 tons, from which 357,634 ounces of gold (0.227 ounce per ton) were recovered. Proven reserves as of January 1, 1974, were 9,054,350 tons with an estimated grade of 0.249 ounces per ton. Total reserves—both probable and possible—were 17.8 million tons.

Homestake has been diversifying. It has a 30-percent interest in the United Nuclear-Homestake Partners uranium operation that recovered more than one million pounds of uranium oxide in 1973. A wholly owned subsidiary (Homestake Forest Products) owns 60,000 acres of timberland in South Dakota and Wyoming. And the firm also has a 57-percent interest in the Madrigal mine in Peru that produces copper, lead, and zinc.

Unofficial estimates indicate that Homestake's 1974 volume will reach $130 million, while earnings could approximate $3.15 per share, up from 1973's $2.00 per share. Dividends have been paid since 1878, except for 1943-45.

DOME MINES LTD.

Dome Mines Ltd. owns and operates a relatively high-cost gold mine in the Porcupine District of Ontario. The Dome mine produced 148,512 ounces of gold in 1973 from 682,200 tons of ore assaying 0.218 ounce per ton. Dome received an average price of $97.32 per ounce for its output in 1973, which was sold on the open market; the 1974 figure was higher.

Although its own mine is classified as high cost, Dome has a 56.8-percent interest in Campbell Red Lake Mines which operates a low-cost mine with large indicated reserves. It also has a 19-percent interest in Sigma Mines, a smaller operation in Quebec. (Sigma produced 78,203 ounces of gold in 1973 from 521,000 tons of ore grading about 0.15 ounce per ton.) A wholly owned subsidiary, Dome Exploration, conducts exploration programs on behalf of Dome, as well as Campbell Red Lake and Sigma Mines.

Dome also has a sizable stake in the petroleum field through its 21-percent interest in Dome Petroleum Ltd. Other holdings are in Mattagami Lake Mines and Canadian Tungsten Mining.

Dome's earnings for 1974 have been tentatively estimated at $3.50 per share compared with $2.52 in 1973 (adjusted for a 3-for-1 split). The company has a moderate capitalization (5.8 million shares) with no priorities ahead of the common stock, as well as an exceptionally strong financial position. Dividends have been paid since 1920. These factors, as well as the fact that its oil and gas income acts as a cushion against the swings in gold prices, makes the shares an investment-quality gold equity.

CAMPBELL RED LAKE MINES LTD.

Campbell Red Lake Mines is a leading Canadian gold producer with an above-average record. The debt-free company, which initiated dividends in 1952, has a low-cost mine in the Red Lake District of western Ontario. In 1973 the company milled 303,976 tons of ore averaging 0.6457 ounces per ton. The average price received was $97.24 an ounce compared with $56.22 (Canadian dollars) an ounce in 1972.

Campbell Red Lake is participating in a mining exploration program with Dome Mines Ltd. (which controls 56.8 percent of Campbell Red Lake), Sigma Mines, and Dome Petroleum Ltd.

Although Campbell Red Lake's 1974 gold production isn't believed to have differed greatly from the previous year, quotations for gold (which averaged around $100 per ounce in 1973) were substantially greater in 1974 and this probably boosted '74 revenues to around $30 million as compared with just under $20 million in 1973.

Initial earnings estimates for 1974 were in the neighborhood of $1.60 per share, up from the $1.11 a share (adjusted for a 2-for-1 split) realized in 1973.

The long-term outlook for the company is enhanced by its large reserves of relatively high-grade ore. However the stock has traditionally been quite volatile, reflecting the fact that some 4.5 million shares out of a total of nearly 8 million are held by Dome Mines.

AGNICO-EAGLE MINES LTD.

Agnico-Eagle Mines Ltd. (NASDAQ over-the-counter quotation symbol: AEAGF) has been operating several silver mines in eastern Ontario, Canada. Since 1953 the silver division has averaged production of slightly less than one million ounces per year.

For some time the company has been focusing on a move into gold production in Joutel Township, Quebec. In 1973 the firm completed a mill with a planned 1,000-tons-per-day ore production rate. However normal delays were encountered in implementing this goal, though the company reported that such equipment was expected to be operational in order to expand gold production to some 100,000 ounces annually by late 1974 or early 1975.

Proven and probable ore reserves at the end of 1973 amounted to just under 765,000 tons averaging an undiluted grade of 0.353 ounce of gold per ton. Possible or drill-indicated reserves totaled 1,947,774 tons averaging an undiluted grade of 0.320 ounce per ton.

Agnico-Eagle, which has 13.8 million shares outstanding, has an excellent long-term potential for an investor willing to accept the normal vicissitudes of a new mill operation.

Chapter 9

Silver, Platinum and Palladium

"The budget should be balanced, the Treasury should be refilled, public debt should be reduced, the arrogance of officialdom should be tempered and controlled, and the assistance to foreign lands should be curtailed lest Rome become bankrupt. The mobs should be forced to work and not depend on government for substance."

—Cicero,
addressing the
Roman Senate

If you play the game of word association and you mention coffee, the average person will say tea. Mention gold and silver comes to mind.

This isn't surprising for silver, like gold, has a colorful history, has performed monetary functions, and has important industrial uses. However, unlike gold it has been successfully downgraded as a monetary metal.

To put it in perspective, it might be said that gold is the star player in the pageant of metals; silver is one of the supporting cast. As some professionals prefer to look at it, silver holds an intermediate position between gold—the prime monetary metal—and copper—the prime industrial metal.

Although silver is unlikely to acquire gold's lustre, it can be a dazzling commodity at times. Early in 1974 it skyrocketed in price to record high levels of better than $6.40 an ounce (spot), approximately doubling in about two months' time. Then it experienced a sharp reaction, losing $1.65 an ounce in the space of a week.

The factors affecting silver prices are similar to those influencing the gold market—diminished confidence in paper currencies, concern over inflation, political and economic turmoil, and so forth. However the sharp escalation of prices in early '74 also reflected other considerations.

One of these extraordinary factors was the widespread rumor that a Texas millionaire had a huge speculative position in silver, reportedly amounting to somewhere in the neighborhood of $200 million.

Moreover, at the time of silver's surge the possibility that Americans might once again be permitted to own gold seemed remote—and in some quarters silver is looked upon as an alternative to gold. It was a way to hedge against inflation (i.e., depreciating currencies).

All of these elements, plus the tightening noose on the free world's energy supply by the OPEC countries, helped push silver above $6 an ounce.

These stratospheric prices soon proved untenable. With the lifting of the Arab oil embargo, silver cascaded downward. Just as there had been some seventeen days in a row in which silver recorded maximum daily price gains in the month of February, there were at least six consecutive trading sessions in March in which the metal fell by the maximum 20¢-per-ounce limit.

Silver's sensitivity to news events and to emotions, rather than statistics, is, of course well known. In 1967-68, for example, the metal climbed from less than $1.30 an ounce to $2.50, then slumped to $1.29 in 1971 while speculators were looking for prices to climb to $4 or $5 an ounce. Treasury officials also have commented on the fact that the price of silver has managed to climb at times when the government was a heavy seller of the metal; then, paradoxically, the price collapsed after the government halted its sales.

Gyrations in price, often unforeseen and unpredictable, must be regarded as one of the characteristics of the silver market. However it is often helpful to seek to balance the forces affecting the price of silver in order to assess its future price possibilities.

The influences pointing in the direction of higher silver prices may be summarized as follows:

1. The deterioration of the international financial and monetary picture suggests a continued stampede into all valuable commodities, of which silver is a prime example.

2. Strength in world gold prices usually has a constructive impact on silver, its "sister" metal.

3. The market has already absorbed most of the large private U.S. holdings of the metal that were built up until 1970.

4. Statistics show a consistent imbalance between production and use of silver. In 1973, for example, new silver production satisfied only about half of the world's industrial demand. Most of the remainder came out of inventories, reducing available stocks from 924 million ounces to 760 million ounces by the end of the year.

5. Some markets for silver are showing impressive growth. One example is the burgeoning business in commemorative and collector arts where silver consumption doubled from 11.5 million ounces in 1972 to 23 million ounces in 1973. The "medal mania" that has captured the fancy of the American public is reflected in the striking of medals honoring just about everyone from the great to the obscure, and from signers of the Declaration of Independence to Hollywood stars.

Are medal collectors really art fanciers or just silver speculators? It is difficult to say, but when one considers that an estimated ten million Americans collect these medals, it can hardly be dismissed as an inconsequential fad. As a footnote, however, we would add that while the medallions seem to satisfy the collecting instincts of millions and offer an interesting pastime, they are a decidedly poor method of hoarding or investing in silver. The silver content in most cases is such that the price of the metal would have to reach astronomical levels to justify a purchase on that basis alone.

6. The conventional industrial uses of silver—notably photography and the electrical-electronics field that together account for roughly half of all U.S. silver consumption—provide assured outlets for the metal. Other significant industrial markets are brazing alloys, electroplating wire, and jewelry.

The chief depressants that tend to point to lower silver prices in the future are these:

1. The U.S. Government still has substantial quantities of the metal in its stockpile and it could release much of this for sale.

2. Minting of coins with a high silver content has largely ceased worldwide.

3. During recessionary periods, silver prices tend to weaken because of reduced industrial demand. (Government statistics indicated a decline in consumption of about 5 percent in the first half of 1974.)

4. Higher prices for silver tend to shrink industrial demand—for instance, big industrial users are likely to stay out of the market when silver crosses $6 an ounce.

5. More than 65 percent of newly mined silver comes as a by-product of copper, zinc, lead, and other base metal deposits. In effect this means that the supply and demand for the base metals influence silver production more than

the supply and demand for silver itself. The great bulk of any new silver production, therefore, would have to come from increased base metal operations. Demand for base metals is growing fast. Ergo, more silver on the market.

6. High silver prices can draw substantial quantities of the metal from hoarders, notably from India. In fact, with the Indian Government's legalization of silver exports in February 1974, silver's price eased as speculators began to ponder the potential of a huge outpouring of the metal from the subcontinent. It has been reported that between three and five billion ounces of silver (mostly in the form of bracelets, earrings, and religious objects) exist in India. During 1973 alone some twenty-six million ounces were smuggled out of India and Pakistan, roughly quadrupling the estimated amount of the metal leaving those countries in the previous year.

A Volatile Metal

Withal, it may be concluded that wide price swings—frequently dictated by emotional and psychological factors—will continue in the silver market. Merely consider that in 1964 a teaspoon in sterling contained $1.29 worth of silver. In February 1974 the same teaspoon, weighing about one ounce, had $6.70 worth of silver; by the end of September '74 the silver in the same spoon had dropped in value to $4.29.

The volatility of silver has caused some industries—notably photography—to devote considerable effort to finding a cheap silver substitute. Whether this eventually comes about is a moot question; in the foreseeable future, it can be argued that silver accounts for only a small portion of the material costs of photography-oriented companies and therefore, for these firms, price swings in the metal are not as significant as they might seem to be.

In general it is our conclusion that while silver is scarcely on a par with gold as a haven against currency inflation and "Lazyfare" economics, and involves an appreciable degree of price risk, the metal cannot and will not be ignored.

During periods of market weakness, opportunities will arise to take silver positions, but those unfamiliar with trading techniques should be sure to avail themselves of competent advice before dabbling in this volatile market.

For the speculative-minded, the silver futures markets in New York, Chicago, and London offer the biggest potential rewards, with commensurate risks. A typical transaction on the Chicago Board of Trade, for example, could generate a 62-percent profit on a 5-percent rise in silver prices but, of course, this type of leverage also works in reverse.

Direct purchases of silver also may be made through banks or brokers. The silver is usually sold in bricks with officially listed brands or markings. A 1,000-ounce brick is about the size of a loaf of bread and weighs about 75 pounds; smaller bricks of 100, 10, and even one ounce also are available. In view of storage problems and insurance costs, we would advise against taking physical possession of large quantities of silver.

SILVER PRICES
(New York)

	High	*Low*	*Average*
1974*	$6.70	$3.27	$4.73
1973**	3.28	1.96	2.56
1972	2.05	1.39	1.68
1971	1.75	1.29	1.55
1970	1.93	1.57	1.77
1969	2.03	1.54	1.79
1968	2.57	1.81	2.14
1967	2.17	1.29	1.55
1966	1.29	1.29	1.29
1965	1.29	1.29	1.29
1964	1.29	1.29	1.29

Source: Handy & Harman

*Through September.

**The general price freeze in effect forced suspension of Handy & Harman's daily quotation for 22 days during July and August.

Silver Bullion Bar

WORLD SILVER CONSUMPTION
(Excluding Communist Areas)
(in millions of ounces)

	1973	1972	1971	1970	1969
Industrial Uses:					
United States	190.0	151.1	129.1	128.4	141.5
Canada	8.5	7.4	6.0	6.0	5.7
Mexico	11.5	6.0	5.1	5.4	4.4
United Kingdom	31.5	27.5	25.0	25.0	24.5
France	22.5	20.0	15.6	15.5	19.3
West Germany	60.0	60.0	59.9	48.2	50.0
Italy	33.5	32.0	30.5	32.0	29.0
Japan	67.5	54.3	46.5	46.0	41.5
India	13.0	13.0	16.0	16.0	16.0
Other Countries	25.0	20.0	17.7	16.4	18.7
Total Industrial Uses	463.0	391.3	351.4	338.9	350.6
Coinage:					
United States	1.5	2.3	2.5	.7	19.4
Canada	1.5	.1	.2	—	—
Austria	6.0	6.3	4.2	4.1	3.3
France	1.0	.8	.4	3.7	.6
West Germany	6.5	24.0	17.9	7.3	5.0
Other Countries	3.5	3.0	2.0	11.1	11.7
Total Coinage	20.0	36.5	27.2	26.9	40.0
Total Consumption	483.0	427.8	378.6	365.8	390.6

Note: Figures for 1973 are preliminary.

Source: Handy & Harman

Perennial Problems

A few final words about silver.

There are some perennial problems faced by the silver speculator that should be recognized.

One is the uncertain reliability of much of the data that is available concerning the metal. Although Handy & Harman, the world's largest silver dealer, and the Bureau of Mines provide a great volume of useful statistics, fundamental data on silver for much of the world is of dubious quality. This can lead to erroneous and conflicting conclusions, depending upon which source of information you rely. Also some bullion dealers have a bias in favor of their prime customers (industrial), who naturally favor lower prices. So treat their comments with suspicion.

Secondly we would caution against any rule-of-thumb approaches to analyzing the silver market. For instance the so-called silver-gold ratio has practically no value as a forecasting tool despite assertions to the contrary. The ratio of about 37-to-1 in favor of gold that existed in September 1974, for example, might be considered to be relatively high in comparison with the 20- or 30-to-1 ratios that have existed for a large part of the past two decades. But we would apply no particular significance to this ratio as it is too susceptible to statistical aberration. It is, in any case, not a timing mechanism—and thus useless.

Finally we would caution against placing silver in the same class as gold as a hedge against political and fiscal irresponsibility. Silver has great charisma, no question about that, but, unfortunately, it is also a highly volatile commodity with a limited role to play as a monetary metal.

The Place of Platinum

Many people interested in gold and silver also find platinum and, to a lesser extent, palladium captivating. A brief mention of these latter metals is in order.

Platinum, a whitish, steel-gray element, has a high melting point, exceptional resistance to corrosion, and—perhaps most important in terms of today's market—exceptional catalytic properties.

Platinum began to find favor in the early 1900s when it was employed as a lining for chemical vessels and in various other industrial uses. In the '20s, International Nickel began to produce platinum as a by-product of nickel, and this opened new vistas for the metal.

A major breakthrough occurred in the '30s when a platinum-rhodium catalyst was utilized in producing nitric acid. Another breakthrough came with the development of platinum nozzles containing microscopic holes for use in spinning synthetic fibers.

Later on platinum began to assume a significant role in the automotive field. With the oil industry eyeing higher octane fuels, researchers found that a platinum catalyst was capable of fulfilling a vital role in the reforming process.

Today the electrical, chemical, and jewelry industries are important sources of industrial demand, but it appears likely that the auto and petroleum industries hold the key to the metal's future in the period ahead. These industries, autos and petroleum, in all likelihood will account for 35 percent of world demand for platinum in 1976.

The most promising major new area of usage is in catalytic converters for autos. Impala Platinum, the world's second-largest producer of the metal, has stated that "demand by the automobile industry for platinum (and palladium) for catalytic converters, will, for some years, provide a relatively stabilizing element in the market for these metals which in the past has been notable for wide fluctuations."

It has been estimated that by the end of 1976 world demand for platinum may expand more than 60 percent over the 2,079,000 ounces indicated for 1973. By far the biggest source of supply is South Africa, which, it is estimated, will produce 1.8 million ounces in 1974, followed by Russia with 780,000 ounces, and Canada with 187,000 ounces. Far down the list are Colombia, the United States, and Japan which together will account for an estimated minuscule 50,000 ounces of the metal. It should be noted that the output of Canada, like that of the Soviet Union, is largely geared to nickel-copper production.

To sum up, it would appear that while the demand outlook for platinum seems favorable, the supply is high. In addition to the normal working inventories in the hands of fabricators and end-users, substantial stocks are already held by investors and speculators, including some who look upon the metal as offering a hedge against inflation. Thus, although the metal will offer trading opportunities from time to time, it should be remembered that substantial increases in productive capacity are likely to keep a lid on platinum prices. And thus caution should be the watchword for would-be speculators in this metal.

The Palladium Picture

Palladium—with platinum, rhodium, iridium, ruthenium, and osmium—is a member of the so-called "platinum group" of metals. Palladium is silver-white, malleable, and ductile and does not tarnish in the air. Palladium is harder than platinum, for which it sometimes serves as a substitute.

Some industries substitute palladium for gold, depending upon relative prices of the two metals. In fact palladium (which is volumetrically greater than gold) tends to be rather closely correlated in price with the yellow metal, whereas the price of platinum tends to have a kindred relationship with silver.

The principal markets for palladium are in the electrical, chemical, dental and medical, jewelry, and decorative industries.

Palladium is an interesting precious metal, but from an investment or speculative point of view it is entitled only to comparatively minor attention in comparison with gold and/or silver.

Chapter 10

Conclusions, Recommendations, Predictions

"Between trusting in the natural stability of gold and the natural stability of the honesty and intelligence of the members of government, I advise you to vote for gold."

—*George Bernard Shaw*

By late 1974 consumer confidence in the United States sank to the lowest point in the twenty-eight year period that the University of Michigan's Survey Research Center has been sampling consumer attitudes. The survey, using February 1966 as a base of 100, reported an index of only 64.5 percent in the most recent poll.

The University of Michigan team reported that "there isn't any reason to believe that the current decline is temporary," but rather that "experience suggests that the present combination of very deep consumer pessimism and decline in real incomes might make for a severe recession."

The Survey went on to point out that an "inflation psychology" prevailed, which should come as no surprise to anyone who has monitored the continuing double-digit inflation rate in the United States. Growing unemployment in key industries such as autos, a catastrophic decline in housing starts, and concern over a possible liquidity squeeze have spread fear both in Wall Street and Main Street.

Of course, we are by no means alone in experiencing difficult times. In Great Britain, the mother of parliamentary democracy, the economic scene is truly horrendous, with labor and capital at each other's throats. Italy apparently is in even worse shape, teetering on the brink of bankruptcy.

Where will it all end?

Some commentators see another Great Depression, worse than that of the '30s, with "blood running in the streets." In this grim scenario, currencies, bonds, stocks, almost any type of asset in fact, will become worthless. The world's economies will grind to a halt and people will have no choice but to "run for the hills" to a Thoreau-esque retreat.

We do not see things in quite that light.

Despite their already overextended positions, it cannot be said that the world's central banks are powerless. They can and will move to meet crisis situations—as exemplified by Germany's loan to Italy. If they open up the monetary spigot and reinflate, they could keep things going for quite a while, though this would merely be buying time at the cost of currency debasement.

There is a saying in Wall Street that "the Fed writes the market letter" and it is undeniably true. The Federal Reserve System, after all, determines how much General Motors or A.T. & T. have to pay to float their bonds, as well as the interest rate individuals must accept on their home mortgages. With that kind of clout, the Fed can slow the economy further or stimulate it.

Which course will it take? If the past is any guide, there is no question that inflation—not deflation—is the most politically acceptable mechanism to deal with the painful process of economic adjustment. Furthermore, new social legislation requires reflation and thus more demand is built into law.

Who can doubt that politicians will not follow the exhortations of the Keynesian economists and opt for more inflation?

In this connection we find these observations of 1974 Nobel prize-winning economist Friedrich von Hayek extraordinarily illuminating:

> The responsibility for current worldwide inflation, I am sorry to say, rests

> wholly and squarely with the economists, or at least with that great majority of my fellow economists who have embraced the teachings of Lord Keynes . . .
>
> It was on his advice and even urging of his pupils that governments everywhere have financed increasing parts of their expenditure by creating money on a scale which every reputable economist before Keynes would have predicted would cause precisely the sort of inflation we have got. They did this in the erroneous belief that this was both a necessary and lastingly effective method of securing full employment.
>
> The seductive doctrine that a government deficit, as long as unemployment existed, was not only innocuous but even meritorious was of course most welcome to politicians. The advocates of this policy have long maintained that an increase of total expenditure which still led to an increase of employment could not be regarded as inflation at all.[1]

We must, of course, take note of the changing lineup of Free World political leaders. Within a matter of months, Israel's Golda Meir, France's Georges Pompidou, Germany's Willy Brandt, Japan's Kakuei Tanaka, and America's Richard Nixon all departed from the political stage.

Many are not optimistic about their successors who will be dealing with inflated-related problems: especially Helmut Schmidt of West Germany, France's Valery Giscard d'Estaing, and America's Gerald Ford. But, on the other hand, one of the most striking failures of the Western World in recent times has been its inability to produce outstanding leaders in the Churchillian mold. Change does not necessarily beget improvement, but somewhere in the current crop of political hacks there could be some talented people. Let's hope so.

In our overall investment strategy we take leave of some of our contemporaries, to whom gold is the ultimate answer to all problems. We have made gold a cornerstone of investment policy, not a religion.

In our view gold is a monetary commodity with universal attraction from an intrinsic as well as a psychological standpoint. As Janos Fekete, Vice-Chairman of the National Bank of Hungary, has remarked, there are about 300 economists in the world who are against gold, but about 3 billion inhabitants of the world who believe in gold.

We concur. We also are convinced that the failure to return to a system in which there is some provision for gold convertibility, plus control over the creation of paper currency, dooms the world to continuing inflation. And inflation, in the words of Pulitzer Prize winner William Caldwell, "has put poverty within reach of all of us."

Possible Remedies

There are some specific measures we would propose to ameliorate our economic ills and set the stage for a return to a sound dollar, which would, in turn, help other currencies:

1. Following an increase in the official price of gold to a figure moderately above the free market price, there should be partial convertibility of gold into dollars.

[1] *New York Times*, November 15, 1974.

2. There should be a legal restrike of a gold coin in the United States.

3. We should consider a 5-percent band in which currencies could fluctuate 2½ percent up and 2½ percent down.

4. We should also face the task of reversing the influences that led to runaway inflation by reducing government spending and balancing the budget.

5. The rate of increase in the money supply should be severely restricted.

6. Despite the urgings of those who press the panic button when unemployment creeps up past 5 percent, we should strenuously avoid taking any measures that will tend to heat up the economy.

We would also recommend that steps be taken toward adopting some form of indexation—the system that links prices, wages, and interest rates to a measurement of the overall price level. It is, in simple terms, an adjustment for inflation. Since politicians simply will not "bite the bullet" in combating inflation, the only solution is to provide some method that adjusts both prices and wages to an overall inflation measurement.

Indexation has worked well in Brazil where capital previously was virtually unobtainable. Belgium has experimented with it and France's Giscard d'Estaing has suggested that it could be used to guarantee the revenues of oil producers.

Indexation actually is already a fixture in many areas of the U.S. economy (e.g., many wage contracts are tied to the cost of living) and we are convinced that it is here to stay whether we like it or not. We don't, but that's incidental to the realities of the situation.

Gold's Potential

Inasmuch as we emphasize gold in our investment strategy, we will attempt to delineate our views on future market possibilities for the metal.

There have been many widely varying predictions as to future price levels for gold—anywhere from $150 to $1,000 per ounce has been forecast, depending upon the time frame for the projection. Between these extremes, Dr. Nico Diederichs, South Africa's Finance Minister, has calculated that, based on an eightfold appreciation of all commodities since 1934, gold should command a price of around $280 an ounce.

Even if we were to assume that gold is just a commodity (which it isn't), it would still have some distance to go to reflect four decades of inflation. The Reuters commodity index has gone up a multiple of fourteen since 1930. We have calculated that from $20.67 an ounce in 1930, it would require a gold price near $290 today just to reflect commodity price increases.

Of course gold is not just a commodity but a universal yardstick of value. In the United States the money supply (demand deposits, currency in circulation, and time deposits) is around $600 billion. This dwarfs by 50-to-1 the official U.S. gold supply at $42.22 an ounce, and even represents a 14-to-1 ratio with gold valued at $150 an ounce.

Clearly we may discard the official U.S. Treasury price of $42.22 per ounce as an anachronism. We agree with France's Giscard d'Estaing, who has called for the abolition of the "absurd fiction" of the official price for gold and has also urged agreement by the International Monetary Fund to allow central banks to buy, sell, and count gold reserves "near the realistic market price."

Gold Price Possibilities

A logical base for future projections is the $120-per-ounce price used in the Italian-German loan agreement, which might be considered a new "floor price" for gold.

From our studies we extrapolate a short-term (through 1975) trading range of $150 to $228 per ounce. And a longer-term (through mid-1977) price band of $120-$309 per ounce. We deem this conservative and would rather err in falling short of the top price than overestimating—which is less useful to investors. Also note that our range has a bottom as well as a top and that means both are possible alternatives.

In our price equation, we hasten to add, we have not factored in any expected demand for gold on the part of the oil-enriched Arab nations. They traditionally have been traders rather than holders of gold and this policy may remain—but it is not inconceivable that some day they may demand gold for their oil instead of dollars, or convert part of their reserves to gold.

Ultimately, there is no true potential ceiling for gold. It will go as high as paper money goes low. Or, to put it another way, paper money's value is the value of a politician's promise; gold's value is protected by the inability of politicians to manufacture it.

How much gold will be sold to Americans during the first year of legal trading?

T. deJongh, Governor of South Africa's central bank, has offered as his estimate for 1975 a range of anywhere from one to eight million ounces. This would amount to between $200 million and $1.6 billion at a price of $200 an ounce.

But perhaps even more significant than this suggested range is Governor deJongh's statement that South Africa would not feel any compelling need to sell its gold output if its international accounts showed improvement. Since South Africa is the chief supplier to the free bullion market, any reduction in its offerings obviously would tend to produce higher gold prices.

Although we do not care to enter this guessing game, we feel that every portfolio should have a stake in gold as the ultimate hedge against adversity. We anticipate substantial long-term price appreciation in gold; but even if this forecast does not materialize, the following could be the considerations of an intelligent investor. Assume that the gold-oriented portion of an investment portfolio is 25 percent and the remaining 75 percent is allocated to conventional income-producing media. If our worst fears for the future turn out to be correct, the gold portion would appreciate to an incalculable value; if disaster doesn't strike, then the remaining 75 percent of the portfolio should appreciate more than sufficiently to offset any lack of performance in the gold segment.

Portfolio Ratios

The allocations in the accompanying table are suggested for portfolios of $10,000, $100,000, $500,000 and $1 million and over. The actual percentages, of course, can be tailored to the needs of individual investors in each classification. For instance, those investors who require more in-

SUGGESTED RATIOS FOR GOLD-ORIENTED PORTIFOLIOS

	$10,000 Portfolio	*$100,000 Portfolio*	*$500,000 Portfolio*	*$1 Million Portfolio*
Demand Deposits	15%	5%	5%	5%
Eurocurrency Time Dep.	35%	20%	25%	30%
Gold Equities*	30%	40%	40%	35%
Gold Coin & Bullion	10%	15%	15%	10%
U.S. Treasury Bills	10%	10%	5%	5%
Gold Forward Contracts	0%	10%	10%	15%

*As a general rule, we favor approximately 75% of this portfolio component in South African golds and 25% in North American golds.

$10,000 Portfolio

Demand Deposit:	
Swiss Franc current account (checking-type account)	$1,500
Eurocurrency Time Deposit:	
Six-month Swiss franc certificate of deposit	3,500
Gold Equities:	
Evenly distributed between: West Driesfontein Gold Mining Ltd. St. Helena Gold Mines Limited Homestake Mining Company	3,000
Gold Coin & Bullion:	
Austrian 100-Corona	1,000
U.S. Treasury Bills:	
90-day U.S. Treasury Bill dollar investment	1,000

come than the recommended portfolio divisions would generate can allocate larger commitments in high-yielding South African Kaffirs.

The size of the individual portfolio has a direct bearing on the manner in which the funds are allocated. For instance a lower-bracket investor should concentrate on coins and (to a lesser extent) bullion while staying out of the futures market entirely.

Generally speaking we do not recommend gold futures for anyone with a net worth of less than $100,000 and even then this high-risk area should account for a relatively minor share of the total portfolio.

$100,000 Portfolio

Demand Deposit:	
Swiss franc current account (checking-type account)	$5,000
Eurocurrency Time Deposits:	
3-month Swiss franc int'l certf of deposit	10,000
3-month Dutch guilder int'l certf of deposit	5,000
3-month Deutschemark int'l certf of deposit	5,000
Gold Equities (equal amounts):	
Homestake Mining Company Campbell Red Lake Mines Agnico-Eagle Mines Limited	10,000
Randfontein Estates Gold Mining Limited West Driesfontein Gold Mining Limited Free States Geduld Mines Limited St. Helena Gold Mines Limited	30,000
Gold Coin & Bullion:	
South African Krugerrand coin Austrian 100-Corona coin	5,000
Gold bullion stored in Switzerland	10,000
U.S. Treasury Bills:	
90-day bills (dollar investment)	10,000
Gold Forward Contracts:	
Trading position covering potential commitment on $100,000 forward contract (Comex Exchange)	10,000

A few final words of caution: Implementation of these portfolio ratios should be in accord with the principles outlined in the individual chapters on coins, bullion, gold futures, etc. And it goes without saying that anyone with a sizable portfolio should seek out personalized professional guidance to achieve his overall investment objectives.

$500,000 Portfolio

Demand Deposit:	
Swiss franc current account (checking-type account)	$25,000
Eurocurrency Time Deposits:	
3-month U.S. dollar int'l certf of deposit	25,000
3-month Swiss franc int'l certf of deposit	25,000
3-month Deutschemark int'l certf of deposit	25,000
3-month Dutch guilder int'l certf of deposit	25,000
Gold Equities:	
Homestake Mining Company Campbell Red Lake Mines Limited Dome Mines Limited or Sigma or Camflo Agnico-Eagle Mines Limited	50,000
Randfontein Estates Gold Mining Limited West Driesfontein Gold Mining Limited East Driesfontein Gold Mining Limited Free States Geduld Mines Limited St. Helena Gold Mines Limited	165,000
Witwatersrand Nigel Gold Mining Limited Loraine Gold Mines Limited	10,000
Gold Coins & Bullion:	
South African Krugerrand coin	25,000
Austrian 100-Corona coin	25,000
Gold bullion stored in Switzerland	25,000
U.S. Treasury Bills:	
90-day bills (dollar investment)	25,000
Gold Forward Contracts:	
Trading position covering potential commitment on $500,000 forward contract (Comex Exchange)	50,000

$1,000,000 Portfolio

Demand Deposit:	
Swiss franc current account (checking-type account)	$50,000
Eurocurrency Time Deposits:	
3-month U.S. dollar int'l certf of deposit	75,000
3-month Swiss franc int'l certf of deposit	75,000
3-month Deutschemark int'l certf of deposit	75,000
3-month Dutch guilder int'l certf of deposit	75,000
Gold Equities:	
Homestake Mining Company Campbell Red Lake Mines Limited Dome Mines Limited Rosario Resources Inc. Agnico-Eagle Mines Limited Sigma Mines	100,000
Randfontein Estates Gold Mining Limited West Driesfontein Gold Mining Limited East Driesfontein Gold Mining Limited Free States Geduld Mines Limited St. Helena Gold Mines Limited	150,000
Witwatersrand Nigel Gold Mining Limited Loraine Gold Mines Limited Blyvooruitzicht Gold Mining Limited Harmony Gold Mining Limited	50,000
Gold Coin & Bullion:	
South African Krugerrand coin Austrian 100-Corona coin	25,000
Gold bullion stored in Switzerland	50,000
U.S. Treasury Bills:	
90-day bills (dollar investment)	100,000
Forward Gold Contracts:	
Trading position covering potential commitment on $1,500,000 forward contracts (Comex Exchange)	150,000

Chapter 11

Answers to Twenty Frequently Asked Questions About Gold

When asked about the possibility of a change in the price of gold back in 1958, Prime Minister Harold Macmillan of Great Britain responded:

"That is one of those questions it is even indecent to ask and still more improper to answer."

1. *Why have Americans been barred from owning gold when many other countries were permitting their citizens to own the metal?*

The ban was triggered by the depression of the 1930s when President Franklin D. Roosevelt initiated a series of measures aimed at restoring confidence and prosperity. Roosevelt decided that we must take "control of the gold value of our dollar" and issued an executive order prohibiting private and bank ownership of gold coins, bullion, and gold certificates. The Gold Reserve Act of 1934 made it official. The prohibition against gold ownership persisted thereafter because U.S. Treasury officials took the position that gold should be "demonetized" or, in simple language, relegated to a minor role in our monetary system. President Kennedy tightened the ban in 1961 by ending the right of U.S. citizens to own gold bullion overseas.

2. *Who are the biggest producers of gold?*

South Africa, the Soviet Union, Canada, and the United States in that order. South Africa is far in front with more than 60 percent of estimated 1974 production. The second-ranking producer, Russia, is far behind with 17 percent of estimated output for 1974. Canada's share of production is between 4 percent and 5 percent, while the United States is a relatively minor factor with only about 3 percent.

3. *Who are the biggest owners of gold?*

The United States still has a commanding lead. It owns some $12 billion worth of gold (computed at $42.22 an ounce) or more than double that of West Germany, which in turn is followed by France, Switzerland, Italy, The Netherlands, Belgium, and Portugal.

4. *How much gold is there in Fort Knox?*

About 147 million ounces of the gold owned by the United States are stored in Fort Knox, Kentucky. This is slightly more than half of the 276 million ounces of gold that the United States owns. Gold that is not stored in Fort Knox rests in vaults in New York, San Francisco, Denver, and Philadelphia.

5. *Do other countries permit ownership and trading of gold?*

Yes. Among the many countries permitting their citizens to own gold are West Germany, Japan, Canada, France, Italy, The Netherlands, and Switzerland. The full list is quite long.

6. *What are gold's principal uses?*

The gold supply in the non-Communist world "disappears" into four principal channels. In 1974 private hoarding or investment accounted for some 12.9 million ounces or roughly 32 percent of the total supply; the jewelry industry took 10.9 million ounces or 27 percent; other industrial users aside from jewelry accounted for 9.3 million ounces or 23 percent; and official coinage uses represented 7.2 million ounces or approximately 18 percent.

7. *How can I protect myself against counterfeit gold?*

In several ways. If you take physical possession of gold bullion, insist that it has the assay mark of one of several well-known approved assaying firms. Among them: Engelhard Industries Division of Engelhard Minerals & Chemicals Corp., 430 Mountain Ave., Murray Hill, N.J. 07974; Mathey-Bishop, Inc., Malverne, Pa. 19355; Ledoux & Co., 359 Alfred Ave., Teaneck, N.J. 07666; International Testing Laboratories, Inc., 578 Market St., Newark, N.J. 07105.

If you do not take physical possession, insist on a document attesting to your ownership of the metal and its authenticity. For safeguards in purchasing gold coins see Chapter 6.

8. *Why is there so much hoarding of gold?*

The demand for gold and other precious metals tends to accelerate when there is political, economic, and monetary turmoil. In recent times there has been mounting distrust of paper currencies as a result of runaway inflation throughout the world, including the United States. This was exacerbated by the Arab oil embargo, which raised the spectre of another "crash" that might match the severity of the Great Depression of the '30s. All of this, plus the failure of the major trading nations of the world to agree on a new world monetary system that would lessen the wide swings in currency values and do away with the currency devaluations that have recurred periodically in the postwar period, has intensified the demand for the one standard that seems to retain its value—gold. Gold is "stored sweat" and that is what men want—to be able to put away their earnings to spend at a later date, knowing it will buy as much as when stored. Without that, bank savings accounts and insurance become a mockery.

9. *What do the leading producers of gold do with their output of the metal?*

South Africa's output is largely shipped to the London market where it is sold at the daily "fixings." However it is anticipated that an increasing proportion of South African gold will go into the Krugerrand, a one-ounce gold goin that is expected to absorb nearly 30 percent of new production.

10. *What is the background of the "two tier market" for gold?*

In 1968 demand for gold reached unprecedented proportions and the United States halted sales in the private market. The United States and other Free World countries signed the Smithsonian Agreement that created the so-called two-tier market. This included a monetary tier, which provided that gold would be exchanged among the central banks of the signatory nations at $35 an ounce. The second, or free market, tier was the uncontrolled market where gold transactions could be entered into by speculators, investors, and industrial users. The free market or commercial price is the more significant today because it is the price that speculators and investors are willing to pay for the metal.

11. *What is the difference today between the "free" market price of gold and the "official" U.S. Treasury price?*

The official price of gold is now $42.22 per ounce. It was set by President Nixon in February 1973 when he devalued the dollar for the second time. It is the price at which the Treasury theoretically stood ready to buy and sell gold, but it must be considered an artificial price because no transactions are taking place at this level.

The "free" market price is the price at which actual transactions have been taking place in the world market. The free market price is set in the London Gold Market or in Zurich. The London market was originated in 1919 at the offices of N. M. Rothschild & Sons where market price "fixings" have been held ever since with few interruptions. There are currently five members or dealers in this market, each of whom represents the needs of producers, industry, the arts, speculators, and hoarders. At two meetings each day, which take place at 10:30 A.M. and 3:00 P.M. London Time (5:30 A.M. and 10 A.M. New York time), the dealers arrive at a price for gold based on supply and demand.

12. *What significant changes have occurred in the use of gold in financial dealings in recent years?*

In 1971 the United States suspended the convertibility of gold for dollars held by monetary authorities of foreign countries. In December of the same year the United States raised the official price of gold from $35 an ounce (which had prevailed since 1934) to $38 an ounce, thereby devaluating the dollar. In February 1973 the United States further devalued the dollar by setting the official price of gold at $42.22 an ounce.

A most significant event of 1974 was a loan agreement between West Germany and Italy under which Italy placed its gold holdings as collateral, not at the official price of $42.22 used for central bank dealings, but at the free market price over a period of weeks, which worked out to about $120 an ounce. The price at which this loan was collateralized, $120 an ounce, though below the free market price of gold, is viewed by some as a "floor" below which it is unlikely that any gold-holding nation will surrender any of its reserves of the yellow metal.

13. *Where can I buy gold?*

You are able to purchase gold through such outlets as brokerage firms, banks, department stores, and jewelers. In addition, a number of U.S. commodity exchanges are prepared to trade in gold futures, i.e., gold for future delivery.

14. *Should I "shop around" if I buy gold?*

Absolutely. You should follow the same general rules that apply to almost any commodity or product. Compare prices asked by sellers, and be sure to ask for an itemization of charges for storage, insurance, assaying, etc. This is especially important if you are buying gold in small quantities; in such transactions the commissions will be relatively high and there will be several middlemen taking a profit out of the final price you pay. Be sure to inquire about the procedures that will be required for you to sell the gold back to the original seller.

15. *Should I buy gold bullion, gold coins, gold shares, or trade gold futures?*

It depends upon your needs, interests, and financial capabilities. If you are a small investor, you might consider gold coins and/or gold mining shares; if you have sizable capital to invest and can assume a fair degree of risk, then gold bullion and/or gold futures might serve as trading vehicles. In each instance it should be recognized that substantial risks are involved in buying and selling bullion, and especially in the gold futures market. (These gold-oriented investments are covered in individual chapters of this book.)

16. *Does gold have any drawbacks as an investment?*

The most obvious one, perhaps, is that it produces no income for its owner, who, therefore, is dependent on a rise in price to show a profit. Gold is a highly volatile commodity that will rise or fall in line with changing conditions in the world of international business and finance. Conversely it has a monetary floor price—which other investments do not.

17. *Will gold increase in price?*

It has risen in the free market from $35 in 1971 to $90 an ounce in September 1973, to a record $197.50 in December, 1974, but has since moved lower. During this time the securities markets of the world were generally in a downtrend. Estimates of future price trends vary widely, but all that can be said with certainty is that the price will fluctuate in line with supply and demand.

18. *Is the gold market regulated or unregulated?*

Largely unregulated. However, this may be subject to change in the U.S. There is movement in Congress to set up a Commodity Futures Trading Commission that will be to the commodities field what the Securities & Exchange Commission is to the securities business. The proposed five-member CFTC group would have under its domain such commonly traded commodities as sugar, cocoa, and coffee—as well as silver and, presumably, gold. Commodity trading advisors, incidentally, would also come under the new law. The Securities and Exchange Commission so far has indicated that it does not regard gold sales as being subject to the securities laws.

19. *When is the best time to buy gold?*

That is strictly a matter of judgment. However there is an old and fairly reliable investment maxim to the effect that it pays to buy on weakness and sell on strength. This seems to be a good rule to apply for anyone who is considering an investment or speculative commitment in gold. If you let your emotions rule you, then you will want to do the opposite. Try to resist the urge.

20. *Where should I keep my gold?*

It depends upon the amount of gold you have purchased and your personal desires. Coins, small bars, and wafers may be kept in a safe deposit box. And a few at home. But it would be unwise to take delivery of a standard 400-ounce gold bar or ingot weighing 27½ pounds unless you have exceptional safekeeping facilities. It would be much more convenient for you to instruct the agent who arranged for the purchase to have the gold stored in a commercial bank or other depository. First National City Bank (N.Y.) for example, is a full-service depository, licensed to issue receipts on The Commodity Exchange of New York. The New York Mercantile Exchange and the International Monetary Market of the Chicago Mercantile Exchange. Swiss banks, too, will store gold for you. There are also safe deposit companies that operate independently of banks.

Chapter 12

A Monetary and Foreign Exchange Glossary of Terms

"A gold price adjustment should be looked upon not as a sin, but as a contribution to the achievement of such basic objectives as sustainable economic growth and reasonable freedom in trade and payments."

—*Dr. Miroslav A. Kriz*

Actual—The physical commodity (e.g., gold) as opposed to futures contracts.

ADR—American Depository Receipt issued by an American bank. The bank first buys a security in a foreign company, then issues an ADR to represent the security. Thus whoever buys the ADR has the ultimate claim on the underlying stock. The purpose of issuing ADRs is to simplify the physical handling of securities of foreign issuers, which sometimes can be cumbersome.

Arbitrage—Buying of foreign exchange, securities, or commodities in one market and simultaneously selling in another market. By this technique a profit can be made because of differences in rates of exchange or in prices of securities or commodities.

Balance of Payments—Net difference of all credits and debits from one country to another.

Barter—Exchange of commodities, using value of merchandise as compensation instead of money. This method has been employed in recent years by countries in which currencies are blocked.

Big Three—The term used to refer to the three biggest banks in Switzerland. They are Swiss Bank Corporation, Union Bank of Switzerland, and Swiss Credit Bank.

Big Five—A term used in British banking referring to the five largest commerical banks in Great Britain. These are Midland Bank, Ltd.; Barclay's Bank, Ltd.; National Provincial Bank, Ltd.; Lloyd's Bank, Ltd.; and Westminster Bank, Ltd.

Bullion—Gold in bars or ingots, assaying at least .995 fine.

Central Bank—An agency that has been set up by the government of a country to supervise generally its banking system and its currency, and to act as its alter ego in financial matters.

Commodity—Generally speaking, any raw material, whether it is a mining product or of agricultural origin. A long list of commodities may be traded by the investor, either through leading stock brokerage firms that also trade commodities, or through firms that specialize in commodities. Each commodity exchange usually has specific trading units—that is minimum quantities of a particular commodity that you can trade.

Conversion—Actual exchange of currency of one country for that of another. Such transactions usually pass through banking channels.

Convertibility—Ability of any holder of a currency to exchange it at will and on demand into any other currency or into gold.

Debasement—In modern terminology the reduction in the purchasing power or value of a currency in circulation as the result of constantly increasing prices. Earlier the term referred to a reduction in the precious metal content of gold and silver coins in favor of a greater proportion of lead, zinc, or other nonprecious metals.

Devaluation—Lowering the exchange value of a currency, in terms of others, for any reason, such as external overvaluation. For gold standard currencies, devaluation is effected by first raising the price of gold in terms of the currency. This automatically decreases the number of units of other gold standard currencies required to purchase a given number of units of the first currency. Basic reasons for devaluation are usually overspending by a government.

Dirty Float—A situation that occurs when a government intervenes in the foreign exchange market while usually claiming that it isn't intervening.

Discount—A rate of exchange lower than spot, the spot rate expressed in terms of percentage per annum or points on which a dealer buys or sells foreign exchange for forward delivery.

Eurodollars—U.S. dollars held on deposit with banks outside the United States and used by them to meet their own temporary needs for additional liquidity or to extend loans to commercial borrowers. Loosely, any dollars held by anyone anywhere outside the United States.

Fineness—The purity of gold or silver as a percentage of the total gross wieght. Thus when one says that gold is .995 fine, it is meant that 99.5 percent of the weight is pure gold.

Floating—When central banks do not protect either upper or lower currency fluctuation limits a currency is said to be floating.

Foreign Exchange—A general term applied to transactions between countries or private citizens involving purchases and sales of foreign (other than local) currencies.

Forward Transaction—A transaction in currency which value date is longer than two days after trade date (one day for Canada and Mexico).

Future Exchange Contract—A contract, usually between a bank and its customers, for the purchase or sale of foreign exchange at a fixed rate, with delivery at a specified time. It is generally used when customers want to preclude risks of fluctuations in rates of foreign exchange on funds due them at a future time.

"Gnomes of Zurich"—Figure of speech, often applied humorously to custodians of numbered banking accounts and to foreign exchange dealers in banking houses along the Bahnhofstrasse in Zurich, Switzerland. The phrase is attributed to George Brown, deputy prime minister of Great Britain in 1964, who blamed defects of the English pound on the 'gnomes of Zurich.'

Hedge—Purchase or sale of foreign exchange to protect an asset or liability. This action ensures a fixed dollar rate for conversion of foreign currency.

IMF—The abbreviation for International Monetary Fund, the organization set up at the Bretton Woods, New Hampshire, conference in 1944. The IMF has about 111 members whose duty it is to register their rates of exchange in terms of American dollars and maintain the rates. It was the IMF that began issuing so-called Special Drawing Rights or SDRs back in the 1960s. The SDRs, or "paper gold," were used to settle accounts between countries on the same basis as gold.

Inconvertibility—This is the converse of convertibility and implies simply that any holder of a currency which is inconvertible has no statutory right to demand conversion of that currency into any other currency or gold. Any such conversion is entirely at the discretion of authorities in the countries concerned. Usually conversion in these countries is exercised through some form of exchange control under which applications must be made to authorities for permits.

Inflation—It has been aptly said that a country suffers from inflation of its currency when there is too much money chasing too few goods. Money, by itself, is useless. Its utility lies only in its ability to act as a medium of exchange and purchasing agent. Government budgets that produce excesses of national spending over income (called "deficit financing") are prime inflators of internal currency and credit systems. Inflation actually takes place when more purchasing power (from credits raised by a government) is put into the hands of a community than is withdrawn via taxation. When increases in money supply or spending power are not offset by increases in productivity, resulting in more goods on sale against increases in money supplies, those in possession of excess spending power will use it, paying higher prices for limited amounts of goods offered for sale. This touches off a general rise in price levels, inevitably causing demand for higher wages, and another swirl in a vicious inflationary spiral.

Lot—The basic trading unit in gold coins. For example, a lot contains twenty Mexican 50-peso pieces or twenty U.S. double eagles.

Par of Exchange—An equivalent of a unit of money in one country expressed in the currency of another, using gold as a standard of value (par value).

Premium (in foreign exchange)—Rate of exchange, higher than spot rate, expressed in percentage per annum or points, on which a dealer buys or sells foreign exchange for forward delivery.

Premium (in coins)—The difference between a coin's market value and the value of the metallic content of the coin.

Price Limit—The maximum price move allowed by exchange rules for a given commodity for any one day's trading.

Rate of Exchange—An expression signifying a basis upon which the money of one country will be exchanged for that of another.

Realized Exchange Gain or Loss—A gain or loss resulting from conversion of one currency into another. It often applies to a gain or loss of subsidiary companies abroad, arising from settled or unsettled transactions in foreign currencies, particularly when unsettled balances are in a current working capital position.

Recession—A period of two consecutive calendar quarters when GNP (Gross National Product, the total of all goods and services produced in a nation) declines in terms of 1958 dollars.

Reserve Currency—One currency widely accepted by many nations' central banks in exchange for local currencies. There must be enough confidence in a reserve currency to result in a willingness by these nations to exchange. Until recently the U.S. dollar was an effective world reserve currency.

Revaluation—A somewhat ambiguous term that can mean either a retraction from too great a devaluation of a currency or the adjustment of a condition in which a currency has been undervalued externally.

SDR (Special Drawing Rights)—A means by which existing international reserve assets may be supplemented periodically through a process of international decision, at rates relative to the world's needs.

Security Deposit—A cash amount that must be deposited with the broker for each contract as a guarantee of fulfillment of the futures contract. It is not considered as part payment of purchase. Some exchanges call this margin.

Short—The market position of a futures contract seller whose sale obligates him to deliver the commodity unless he liquidates his contract by an offsetting purchase; also the holder of a short position in the market.

Spread—A market position that is simultaneously long and short equivalent amounts of the same or related commodities. In some markets the term "straddle" is used synonymously.

Straddle—Also known as a "spread," the purchase of one future month against the sale of another future month of the same commodity. A straddle trade is based on a price relationship between the two months and a belief that the "spread," or difference in price between the two contract months, will change sufficiently to make the trade profitable.

Swap—A simultaneous purchase and sale (or sale and purchase) of a foreign currency, mostly against U.S. dollars. This transaction does not protect against risk of parity change. Swaps are a method of funding foreign currency borrowings through a secondary currency.

Troy Ounce—A system of weights in which twelve troy ounces make a pound; gold and silver are measured in troy ounces, which is the system widely used by jewelers in the United States and England.

NOTE

For a more comprehensive glossary we recommend *Financial Tactics & Terms for the Sophisticated International Investor* by Dr. Harry Schultz, Harper & Row, $7.95, 176 pages.

INDEX

INDEX